COMPLEXITY
OF GROUNDWATER
SYSTEMS

Unconventional Subsurface Flow and Transport

Series Editor: Hongbin Zhan (*Texas A&M University*)

Study of groundwater flow and solute transport has been advanced into new territories that is beyond the conventional theories such as Darcy's law and Fick's law in recent decades. The studied media change from permeable porous and fractured ones to much less permeable ones such as clay and shale. The studied pore sizes also change from milli-metres to micro-meters or even nano-meters. The objective of this series is to report recent advances in subsurface flow and solute transport that pushes the knowledge boundary into new territories which include, but not limited to topics in forthcoming titles.

Published

Vol. 1   *Complexity of Groundwater Systems*
         by Teng Ma and Liuzhu Chen

Unconventional Subsurface Flow and Transport – Volume 1

COMPLEXITY
OF GROUNDWATER
SYSTEMS

Teng Ma
Liuzhu Chen

China University of Geosciences, China

W⊕ World Scientific

NEW JERSEY · LONDON · SINGAPORE · BEIJING · SHANGHAI · HONG KONG · TAIPEI · CHENNAI · TOKYO

Published by

World Scientific Publishing Co. Pte. Ltd.

5 Toh Tuck Link, Singapore 596224

USA office: 27 Warren Street, Suite 401-402, Hackensack, NJ 07601

UK office: 57 Shelton Street, Covent Garden, London WC2H 9HE

Library of Congress Cataloging-in-Publication Data

Names: Ma, Teng, author. | Chen, Liuzhu, author.
Title: Complexity of groundwater systems / Teng Ma, Liuzhu Chen,
    China University of Geosciences, China.
Description: Singapore ; Hackensack, NJ : World Scientific, 2023. |
    Series: Unconventional subsurface flow and transport ; volume 1 |
    Includes bibliographical references and index.
Identifiers: LCCN 2022000996 | ISBN 9789811229039 (hardcover) |
    ISBN 9789811229046 (ebook) | ISBN 9789811229053 (ebook other)
Subjects: LCSH: Groundwater. | Groundwater flow.
Classification: LCC GB1003.2 .M3 2023 | DDC 551.49--dc23/eng20220315
LC record available at https://lccn.loc.gov/2022000996

British Library Cataloguing-in-Publication Data

A catalogue record for this book is available from the British Library.

For any available supplementary material, please visit
https://www.worldscientific.com/worldscibooks/10.1142/12064#t=suppl

Desk Editors: Jayanthi Muthuswamy/Steven Patt

Typeset by Stallion Press
Email: enquiries@stallionpress.com

Preface

Complexity science is an emerging discipline dedicated to complexity research that focuses on complex systems. It is the development of systems science. Complexity science requires a major shift in mindset that takes us from a linear, mechanical, and deterministic world to a nonlinear, evolutionary, and unpredictable world. Its methodology is not only a major challenge but also an important supplement to traditional scientific methodology.

The earth is an open, complex, and dynamic giant system. Its fundamental properties determine its complexity and nonlinearity. This requires us to approach the earth system from the perspective of complexity science.

As an important part of the earth system, groundwater exists in various forms in the rock voids and minerals. It is involved in a variety of water–rock interactions under different conditions, driving the material and energy cycle in the earth; thus, a large-scale groundwater system forms with complex structures and processes. The groundwater system provides water, nutrients, and habitat conditions for the earth's ecosystems and plays a pivotal role in the planet's livability.

The groundwater system has a material exchange with the outside; it is a typical "open and complex" giant system. Its complexity is in the system boundary, water-bearing medium, hydrodynamic and hydrochemical fields, etc., and the complexity is not only spatial but also temporal.

This book consists of five chapters, which introduce the basic concepts, theories, and methods in complexity science (Chapter 1), the complexity and influencing factors of groundwater systems (Chapter 2), the uncertainty of observations and prediction in groundwater systems (Chapter 3), research methods of groundwater complexity (Chapter 4), and research examples of groundwater complexity (Chapter 5). Zhiqiang Wang participated in the book's conceptualization. For compilation and revision, Xiancang Wu, Wenhui Liu, and Xiaoli Wang contributed to Chapter 1; Rui Liu, Ziqi Peng, and Zhanqiang Chen contributed to Chapter 2; Wenkai Qiu, Yu Chen, and Mengting Zhang contributed to Chapter 3; Kewen Luo, Chenxuan Shi, and Siyu Jiang contributed to Chapter 4; and Junqi Li, Yanwen Huang, and Qianqian Jiang contributed to Chapter 5. Gongyu Zhou, Ruihua Shang, Yubin Liu, Bolin Zheng, Yaoyao Kong, Wulin Chen, and Saier Chen contributed to the revision.

This book has referenced a large number of works of literature, and the book's authors would like to express sincere thanks to the authors and institutions of those works as well as to all the people involved in the writing of this book. They also thank the editors and the publisher for carefully composing, seriously reviewing, and attentively modifying the manuscript.

About the Authors

Teng Ma is a Professor at the School of Environmental Studies, China University of Geosciences. He devotes himself to teaching and research in the fields of groundwater and the environment. His recent research interests include groundwater–surface water interaction, groundwater pollution and control, water–rock interaction in low-permeability media, and the earth's critical zone in the basin.

Liuzhu Chen is an Associate Professor at the School of Environmental Studies, China University of Geosciences. She is primarily involved with teaching and research on soil–groundwater pollution and control.

Contents

Chapter 1

Complexity Science: From Reductionism to Holism

1.1 Basic Concepts

1.1.1 *Complexity science is an extension and development of systems science and is "the science of the 21st century"*

Complexity science has been a necessary part of modern science since its birth, pointing out another direction for scientific development.

Bertalanffy, a pioneer of systems science, raised the issue of studying complexity in the late 1940s. Weaver, one of the founders of information theory, put forward the division of organized complexity and unorganized complexity in the same period, taking organized complexity as the research object of systems science, which had a profound impact on the subsequent scientific development. However, on the whole, there was no substantial progress in complexity science during this period. In the 1950s and 1960s, the branches of systems science that made important progress were operational research, cybernetics, information theory, and other technical sciences. The research objects belonged to simple systems and had not yet touched on the real complexity. After the 1970s, the theory of simple systems became more and more mature. Systems science turned to take complexity as the main object and tried to establish a general theory of complex systems then. Represented by the establishment of Santa Fe Institute in 1984, complexity science has officially become a popular discipline. What is complexity science? Santa Fe Institute believes the following: "Complexity arises from any system in which multiple agents interact, adapt to each other, and adapt to the environment. These interactions and adaptations produce system evolution processes and common amazing 'emergence' behaviors at the macro level. Complexity science tries to find a common mechanism that can lead to this complexity in such different physical, biological, and social systems".

Complexity science is a new discipline that takes complex systems as its research object and devotes itself to complexity research. A complex system refers to any system made up of multiple subjects interacting, adapting to each other, and the subjects adapting to the environment. Many systems important to humanity exhibit similar complexities, eg. the market is composed of various of buyers and sellers, a multicellular organism is composed of proteins, membranes, organelles, cells and organs, the internet is composed of users, sites, servers, and website, etc. (Holland, 2014). Each of these complex systems shows a unique characteristic called emergence, which is roughly described by a common saying, that is, "the function of the whole is greater than the sum of the functions of each part".

The direct source of the theory of complexity science should be various disciplines of the original systems science, such as general system theory, and various new disciplines, aiming at exploring complex systems such as dissipative structure theory. Therefore, the first source of complexity science is various branches of modern systems science, the second source is various disciplines of nonlinear science, and the third source is various active fields of artificial life research, which are currently popular (see Figure 1.1). The appearance of complexity science has greatly promoted the in-depth development of science, making a human's understanding of objective things rise from linear to nonlinear, from simple equilibrium to non-equilibrium, and from simple reductionism to complex holism. Therefore, we believe that the birth of complexity science marks a brand-new stage in the level of human understanding and will be another new milestone in the history of scientific development.

Previous studies have found that complexity science should not only be reductionist but also surpass reductionism, not only be holistic but also surpass holism. In other words, it is necessary to combine holism and reductionism to form a new methodology suitable for complexity science. Holism in the weak sense and reductionism in the weak sense both absorb each other's advantages

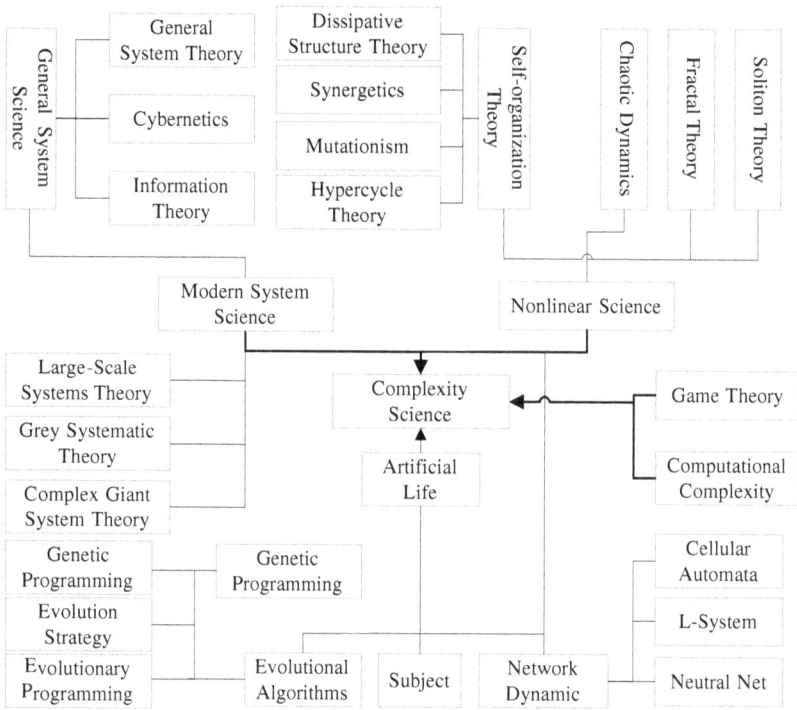

Figure 1.1. Schematic diagram of theoretical composition of complexity science (Huang, 2005).

and overcome their weaknesses, so they are essentially interlinked and have no essential distinction. The theory of integration has become a new methodology of complexity science.

Complexity science is essentially a new form of the overall evolution and development of scientific systems, a new type of science that replaces the dominant position of simplicity (reductionism) science, so it cannot be simply incorporated into systems science and should be regarded as a new stage of the development of systems science. The development of modern science and technology shows that complexity cannot be attributed to the inadequacy of the cognitive process. It must be admitted that there is objective complexity. Even if it has been recognized by people and even if a solution is found, it is still complicated. That is to say, some fundamental differences between simplicity and complexity should be identified, giving complexity science a relatively determined field of study.

With the development of the complexity research field, it has been divided into two sub-fields to study two different complex systems: Complex Adaptive System (CAS) and Complex Physical System (CPS). In complex adaptive systems, adaptability means that the main body of the system has certain autonomous behaviors from a simple conditioned reflex function to more intelligent thinking, learning, and decision-making abilities. Adaptability is the characteristic of a living system. In the life system, the subject is generally a "living creature". There is also a kind of system, which Holland calls a complex physical system. Its main body interaction usually only occurs between adjacent main bodies in space. The main body follows the laws of physics and does not have adaptability (Feynman *et al.*, 2011). Complex systems are often a contrast between simple and complex, relatively simple individuals and unexpected complex group behaviors.

Although still in its budding, complexity science has attracted widespread academic attention, and someone even declared that "the 21st century will be the century of complexity science" (Rycroft and Kash, 1999) and that "the country that has mastered the science of complexity will be the economic, political, and cultural powerhouse of the 21st century". The reason that complexity science has gained such a high reputation is its breakthrough in scientific methodology. The rise of complexity science has had a great impact on traditional scientific methodologies such as reductionism and holism in philosophy. Its research has also adopted many research methods that are seldom used or even excluded by traditional scientific research. Complexity science methodology is not only a major challenge to traditional scientific methodology but also an important supplement to traditional scientific methodology and is of great significance to the healthy development of complexity science itself.

1.1.2   *Systems and complex systems, elements and environments, structures and functions*

1.1.2.1   *Systems and complex systems*

Qian Xuesen, a famous Chinese scholar, believes that a system is an organic whole with specific functions formed by the several interacting and interdependent components, and this organic whole is a component of a larger system to which it belongs (Qian, 2000).

According to Qian Xuesen (Qian *et al.*, 1990), it is possible to classify systems into two main categories of simple systems and giant systems, depending on the number of subsystems and the complexity of the relationship among them. A simple system has less subsystems, and the relationship between them is simple. Some non-living systems, such as a machine, is a small system. If the number of subsystems is relatively large, such as tens or hundreds, for instance, a factory, which can be called a large system. Regardless of it is small system or large system, studying this kind of simple system can start from the interaction between subsystems and directly integrate the motion function of the whole system, and it can use the Newton mechanics method for analysis, but at most it needs to use large computers.

If the number of subsystems is very large, such as tens of thousands, tens of billions, and trillions, it is called a giant system. If there are not many kinds of subsystems in a giant system and the correlation between them is relatively simple, it is called a simple giant system, such as a laser system. Because there are many subsystems, it is impossible to accurately describe the motion of each subsystem. However, due to the simple relationship between subsystems, the details can be omitted and the statistical mechanics method and thermodynamic entropy method can be used to study the system. Dissipative structure and synergetics are theories used to study simple giant systems.

If there are many kinds of subsystems with hierarchical structures, and the correlation between them is very complex, this is a complex giant system. If this system is open, it is called an open complex giant system. For example, a biological system, the human brain system, the human body system, and the geographical system, the ecosystem, the social, the galaxy system, etc. They also have the relationship of inclusion and nesting. These systems are so complex in terms of structure, function, behavior, and evolution that to this day, there are still a lot of problems that we do not know of. If there are still people involved in the system and can learn and adapt, it is a social system. Qian Xuesen thinks that such a system is a special complex giant system. For open complex giant systems, due to the variety of components, the nonlinear interaction between subsystems is extremely complex, and the correlation is nonlinear, uncertain, fuzzy, and dynamic. The system also has a complex hierarchical structure, with time, space, and function nested and influencing each other. There is also the interaction between the system and the environment. The system has initiative, adaptability, and evolution. Therefore, the traditional research methods cannot solve the problem of open complex giant systems, because there is no theory from micro to macro, and there is no statistical mechanics constructed from the interaction of subsystems. Therefore, new research methods and methodologies need to be proposed. For this reason, Qian Xuesen constructively put forward their comprehensive integration method from qualitative to quantitative. The essence is consistent that he and other scholars regard the open complex giant system and the complex system as the research object of complexity science, respectively, but he classifies the complex system more finely.

So, what is a complex system? There are also various views on this. We can understand it from the characteristics of complex systems. The most typical and most accepted view is that Cilliers summarizes 10 aspects of complex systems in his book *Complexity and Postmodernism* (Cilliers, 1998). After generalization and induction, the 10 characteristics can be simplified to the six characteristics: a huge number of components, complex interaction between components, openness, far from balance, path dependence, and distribution parameters.

In a word, from the characteristics of complex systems, complexity science pays attention to the integrity of research objects, the connection between the elements that make up the object system, and the emergence of entirety.

1.1.2.2 *Elements and environment*

Element is the smallest and necessary unit to make up the existence of an objective thing and maintain its movement. It is the fundamental unit of the system, and the driving force for the generation, change, and development of the system. Elements are hierarchical. An element is the element relative to the system in which it is located and a system relative to the elements that it's made of. In the system, they are independent of each other and are proportionally connected into a certain structure, which determines the nature of the system to a large extent. The nature, status, and function of the same element are different in different systems. If one element in the system is too different from other elements, it will be removed by itself.

The environment of the system is the aggregation of various things or conditions outside the system that are related to the existence and evolution of the object system. According to different definitions of the environment, it is one-sided that taking external conditions equal to environment by philosophers in the past. External conditions are not equal to the environment. Firstly, only the aggregate of various external conditions that affect the existence and evolution of the system can be called the environment of the system. A single external condition is just an element or subsystem of the environment of this system. Second, the environment can also be regarded as a system composed of various external conditions. The difference of the boundaries between the environment and the system is relative. For a definite system, its external environment is also relatively definite. On the contrary, when a certain environment is studied as a system, what was originally called the environment is converted into a definite research object system, and the system originally studied is correspondingly converted into a part or element of the above system. It can be seen that the status of the system and the environment can be transformed under certain conditions. A wide variety of systems and environment relationships are studied, with the primary objective of understanding how object systems exist and evolve under the influence of the environment. Therefore, we can refer to the study object system as an ontological system and the environment as an external condition system. Third, as an external condition system, the environment also has all the basic characteristics and attributes of other systems. Moreover, its evolution also follows the general law of system evolution.

1.1.2.3 *Structure and function*

The category of structure and function is a dynamic process that reveals the internal composition of things and the interaction between things and the environment. It is required to considering things from the relationship between structure and function.

There is a certain structure in anything, and the category of the structure is a way to characterize the combination of elements within the thing. The structure has a series of characteristics. First of all, the structure has strong stability and is relatively unchangeable in a thing. Therefore, the structure can be used as a basis to divide various types of things. Secondly, the structure is orderly. In other words, structure, as a way to connect elements, has certain rules. It is manifested in a certain way and is governed by certain regularities. Thirdly, because the structure is both orderly and stable, a development series from a low-level to a high-level and from simple to complex is formed among structures.

The complex of structure is not only limited to the composition of various elements but also the connection of many levels. When hierarchy is introduced into the structure, longitudinal structure and horizontal structure are formed. A horizontal structure refers to how different elements are constituted at the same level, while longitudinal structures reveal how different levels are linked.

Function and structure are different. The function is the ability of things to act on other things, that is, the ability of systems to act on the environment. Function and structure express the same process from different aspects. For thing itself, they refer to what capabilities it has; for other things, they refer to what roles it is. As far as the relationship between things and other things is concerned, it refers to what role it has. The role is exactly how function manifested itself in the interaction of things, and the function of things always manifested in the course of action with the environment.

The function of things depends on many factors, and different hierarchy have different functions. Generally speaking, functions can be divided into the following types:

(1) *Meta-function*: The function of a thing depends first on the nature of the elements that make up. Any element within a thing has a certain degree of relative independence, thus having a certain relatively independent function. The function that an element has in an isolated state without relying, on the whole, is called "meta-function".

(2) *Inherent function*: The function of things depends on the number of elements. Under the premise that the meta-function is not zero, and the other situation is constant, the more the number of elements, the greater the function of the thing. Mechanized sum of meta-function is also called inherent function.

(3) *Structure–function*: The function of things also depends on the structure. The function formed by the structure of things is called a structure function. The function of things depends on both

elements and structures, which is the sum of the "inherent function" generated by elements and the "structural function" generated by structures.

According to the source of function, there are two ways to improve the function of things. The first is to transform the quality of each element and increase the number of elements, that is, to improve the inherent function. The second is to change the structure of things and give full play to the overall role of things, that is, to improve the structural function.

The relationship between structure and function is the unity of opposites. On the one hand, structure and function are different and opposite. The structure points to how features are constituted. Functions point to the relationship between things and the surrounding environment, and the contents to be revealed by the two are different. The difference between structure and function is also shown through the phenomenon of "isomorphic but different functional" and the phenomenon of "same functional but heterogeneous". The phenomenon of isomorphic but different functional refers to the fact that two things with similar structures have different functions. The phenomenon of same functional but heterogeneous means that two things have similar functions, but their structures are fundamentally different. On the other hand, structure and function are interrelated, that is, they interact and transform each other under certain conditions.

The structure and function are unified. First, the structure of things determines the structure function of things. The structure function is a manifestation of the structural ability of things, which can expand or shrink inherent function of things. The structure produces structural function, structure determines function and is the foundation and premise of function. Second, the structure determines the function, but the function is counterproductive to the structure. In the process of exerting the function, things exchange material, energy, and information with the surrounding environment, so that the system absorbs "negative entropy" from the environment. This "negative entropy" increases to a certain extent and exceeds the "entropy increase" of the system itself, then the system would overcome disorder and evolve to an order state. On the other hand, when the "entropy increase" in the system exceeds the "negative entropy", the structure of the system will become unstable and disintegrate, resulting in a state of collapse. Third, the structure and function of things interact and transform each other under certain conditions. New functions would generate when structure of things change to a certain extent, new structures also would generate when function of things play to a certain extent.

1.1.3 *Complexity science, nonlinearity and uncertainty, random theory*

1.1.3.1 *Complexity science*

Complexity science explores the basic laws and applications of the complexity of things from the concept of complexity. Complexity science will provide new ideas, new approaches, new methods, and new applications for research and development in many fields. Specifically, it should include the following contents:

(1) To study the common characteristics, transformation modes, and various mechanisms of complexity science theory and complexity.
(2) To study the manifestation of complexity and its internal laws and their connections such as bifurcation, chaos, turbulence, spatial–temporal pattern and coherent structure, and the "emergence" of various unpredictable phenomena.
(3) To develop the research methods, means, and experimental devices of complexity science.
(4) To study methods and techniques to control complexity, such as chaos control and anti-control, chaos synchronization theory and methods.
(5) To study the application of complexity in many fields.

The group called "complexity science" generally includes the following theories: for example, dissipative structure theory, synergetics, hypercycle theory, catastrophe theory, complex giant system in the modern system science, chaos theory, and fractal theory in theoretical nonlinear science as well as evolutionary programming, genetic algorithm, artificial life, and cellular automata proposed through computer simulation research. This can be regarded as the core of complexity science. At present, the concept of complexity have begun to be applied in various fields of physical science, life science and economic science, and even in fields of humanities and social sciences.

These applications can be seen as the research periphery of the science of complexity. Complexity is not a theory, but a framework, a worldview, epistemology, and methodology. It is a change of epistemological paradigm. The complexity science paradigm is emerging in various fields of natural science, humanities, and social sciences and has gradually formed a complexity paradigm different from the reductionism paradigm (Wu, 2001).

Complexity science is a major shift in the mindset that takes us from a linear, mechanical, and determined world to a nonlinear, evolutionary, and unpredictable new world. Complexity science has some of the following characteristics:

(1) It can only be defined by research methodology, and its measurement scale and framework are non-reductive research methodologies. It is an important feature of complexity science to define its research objects through research methodology.
(2) It is not a specific discipline, but scattered in many disciplines and interrelated. From the traditional discipline to the present interdiscipline, from politics, economy, biology to language, brain, market, and transportation, it is in almost every corner of human life and difficult to tell where its boundaries are. The reason it is called complexity science is that it has a relatively consistent methodological stand beyond reductionism.
(3) It seeks to break the boundaries of traditional disciplines and find a unified mechanism of interlinkage and cooperation among the disciplines.
(4) It should try its best to break the linear theory that has ruled and dominated the world since Newton's mechanics and abandon that reductionism applies to all sciences.
(5) It should create a new theoretical framework system or paradigm and apply a new mode of thinking to understand the problems brought to us by nature.

Complexity science puts forward a new view that complexity arises from the interaction and adaptation between system components and the environment. It emphasizes that complexity is an "emergence" in the system evolution process and has essential unpredictability. However, in terms of research methods, complexity science has parted ways with general system theory and returned to the decomposition and construction of systems. It uses calculation to overcome problems that cannot be solved by mathematical formulas such as huge scale, complex relations, and numerous parameters.

1.1.3.2 *Nonlinearity and uncertainty*

The source of nonlinear science is the traditional system theory. Apart from Qian Xuesen's open complex giant system theory, in essence, it mainly analyses the structure and function of the system, that belongs to static analysis, and it is an organized system without the participation of the organization subject. Since Prigogine put forward the dissipative structure theory in the 1970s, the focus of system theory has changed from existence to evolution. The self-organization theory, which consists of dissipative structure theory, synergetics, catastrophe theory, and hypercycle theory, highlights the initiative and evolution of the system itself. At this stage, complexity science studies the evolution of systems from disorder to order or from one ordered structure to another. The methods used include not only physical experiments or mathematical model analysis but also computer simulation. Therefore, its methodology is non-reductionist. In the aspect of studying evolution, the role of the reduction decomposition method is very limited, because the reduction decomposition method is mainly used to study the composition and structure of substances. The main method used to study evolution is the overall research method, that is, under the condition of keeping the overall integrity of the system, the overall evolution behavior of the system is studied using models, simulations, and other means.

In the 1980s, with the introduction and development of chaos, fractal, and soliton theories, nonlinear science made great progress. At this time, people discovered that dissipative structure theory, synergetics, catastrophe theory, and hypercycle theory also studied nonlinear problems. Chaos, fractal, and soliton theories reveal the rich content brought by system nonlinearity, and the nonlinear interaction of simple elements may produce complex behaviors. Chaos and fractal theory study the complexity and prospect of self-organization from the perspective of time and space, thus they become important theoretical source of complexity science.

The relationship between nonlinearity and complexity is very close. Some people think that complexity must contain nonlinearity, and some even think that complexity is nonlinearity. For example, Klaus Meinzell of Germany wrote this way (Meinzell with

Zeng Guoscreen, 1999): "in the mathematical framework of complex systems in this book, complexity is defined firstly as a nonlinearity, which is a necessary but not sufficient condition for chaos and self-organization. On the other hand, linearity means the principle of superposition. In popular terms, the whole is only the sum of its parts". In complexity science, the often mentioned nonlinear processes, nonlinear phenomena, nonlinear relationships, and nonlinear interactions must also be elucidated by mathematical models or expressions.

There are many kinds of uncertainty in the system, such as randomness, fuzziness, information incompleteness, and ambiguity. Since Newton, science has progressively developed two parallel descriptive frameworks. One is a deterministic description represented by Newton's mechanics, and the other is a probability description developed from statistical mechanics and quantum mechanics. In the early development of system theory, both methods have been widely used, but generally speaking, they are either described by determinism or probability theory, but they do not communicate both. The deterministic theory is used to describe the general system theory, the catastrophe theory, nonlinear dynamics, differential dynamic system, etc. In control theory, operations research, and other disciplines, both descriptions are used, but communication is still not realized by dividing different branches to use them separately. Self-organization theory tries to communicate the two description systems and has made some progress, but it's still not enough. The development of modern science increases calls for the communication of the two descriptive frameworks to form a unified new framework. The development of systems science requires, in particular, to communicate the deterministic framework with the probabilistic framework. The development of new disciplines such as chaos bring a picture of it. One view is that if we take finiteness as the basic starting point for understanding nature and admit the finiteness of nature, we may be freed from the deep-rooted opposition between the two description systems.

1.1.3.3 *Random theory*

Randomness is difficult to define, but it is often associated with disorder, uncertainty, irregularity, and instability. We often use craps to illustrate randomness, but are the results of craps random? Assuming that we have perfect knowledge of dice, force, direction, and other environmental parameters, the result of craps is undoubtedly certain. At this time, randomness is a ignorant description of necessity. Weather forecasts are not always accurate, and their randomness also has ignorance. Randomness is sometimes intrinsic. Take the weather forecast as an example. Lorenz proved the impossibility of its long-term forecast. It is a random feature generated by chaos. Heisenberg's uncertainty principle points out that in the quantum world, we cannot accurately measure the position and velocity of elementary particles at the same time, and its uncertainty obeys a probability distribution. The randomness is also essential.

Research shows that we usually use the concept of randomness in three conditions. First, it means that the state of the thing is very irregular and no law can be found to compress its description. Second, it means that the process of producing the thing is purely accidental or random. And, the results produced by this process are mainly random and its information is incompressible. Third, it refers to the seemingly random results produced by the pseudo-random process, that is, the process is not an accidental deterministic process, but its results are very disordered (such as chaos). In this way, there are two kinds of the randomness, one is process randomness and the other is the result or state randomness. The real meaning of randomness is that not only the results it produces have the characteristics of randomness but also the process is random. Chaos only has the seeming randomness of the result but does not have the randomness of the process.

Objectively speaking, randomness is the interference that occurs in a process. The deviant movement that changes this process is unexpected changes. From the subjective aspect, it indicates the unpredictability and relative indeterminacy of the subject to the object. It is generally believed that randomness is the root of the complexity and even some people equate complexity with randomness. The relationship between complexity and randomness is also very complicated, which requires us to differentiate.

Is it like the more random the system, the more complicated it is? If we speculate intuitively, it seems that this should be the case, but if we analyze it slightly, it seems that this is not entirely the case, depending on which complexity measurement method we use. Due to different measurement methods, two opposing viewpoints have emerged.

The first view holds that "complexity" is equivalent to randomness, and the size of randomness is the measure of complexity (Friedrich, 1993). The more the randomness, the greater the complexity. Completely random information is equivalent to the maximum complexity. Therefore, the maximum complexity is equivalent to the maximum information entropy, and a considerable part of the calculation complexity and algorithm complexity contains this meaning. Entropy, Kolmogorov complexity, and the fundamental complexity defined by Kramer all belong to this kind of complexity.

The second view holds that "complexity" is not equal to randomness, but more profound than randomness (Wu, 2002). Gelman, a Nobel Prize winner in physics, proposed the concept of effective complexity to describe the complexity of science effectively. He used the stories of Shakespeare and the monkey to illustrate that complete randomness and disorder are not the most complex systems. A monkey stands beside a typewriter and assumes that it taps keys randomly. Each time it taps, the probability of any symbol or space bar being taped is equal. Such random knocking is the greatest randomness, but it may not form any characters that are meaningful. On the contrary, Shakespeare also used this typewriter to knock, but he knocked out the immortal Hamlet. Shakespeare's works obviously cannot be random knocking, but meaningful characters, whose randomness is much less than monkey's random knocking. So, is the monkey's random "work" much more complicated than Shakespeare's work? If there is a positive answer, then such complex research has no scientific significance.

In thermodynamics, the internal state of a completely random gas system composed of identical particles is not the random state described by the program, but its result is very simple and definite. Therefore, we cannot get the conclusion that the more random the system is, the more complex it will be.

1.2   Main Features of Complexity Science

There is a variety of perceptions about complexity science, and we can start with the characteristics of complexity science. Here are details on its four main basic features.

1.2.1 *Nonlinearity*

The world is a larger complex system of numerous complex systems, and the essence of complexity science is the study of nonlinearity. Nonlinearity is very closely related to complexity.

Linearity and nonlinearity were originally mathematical concepts, representing the relationship between variables, which can be expressed as a mathematical expression or model. Linearity refers to the proportional relationship between two or more variables. Its function is a straight line in the rectangular coordinate system, otherwise, it belongs to nonlinearity. In physics, a linear system can be described by a linear function. There are three basic properties of linearity, output response, state response, and state transfer, which are all satisfy the overlay principle (Fang, 1996). Therefore, we can only judge whether a thing, relation, or function is linear or nonlinear according to the mathematical relation. It is impossible to talk about linearity or nonlinearity without the mathematical language, at least cannot explain their exact meaning, because words only have their meaning in a certain context.

For systems far from equilibrium, the superposition principle is usually no longer applicable and nonlinearity is inevitable. Because the two solutions of a nonlinear system cannot be combined to get another solution, the nonlinear problem must be solved as a whole.

Although the above definitions of linear or nonlinear systems do not directly use mathematical expressions and are not necessarily equivalent, they show that the linearity or nonlinearity of the system can only be judged by mathematical model, state parameters, or numerical characteristics of the system. Similarly, in complexity science, nonlinear processes, nonlinear phenomena, nonlinear relationships, and nonlinear interactions often mentioned must also be elucidated made through mathematical models or expressions.

Just as the human epistemic process begins with the knowledge of simple things, the natural sciences are created and developed by studying the simple objects of linear systems. For example, in the 16th century, people already knew in practice that the motion trajectory of the projectile was a parabola, but it was impossible to strictly prove why the cannon had the longest range when it was 45 degrees to the horizontal line. Galileo noted that the projectiles move along the parabola due to the combined action of gravity and push, so it can be decomposed into two parts: a uniform linear motion in the horizontal direction and a free-fall motion in the vertical direction so that geometry can explain the problem. It can be seen that the research method of classical natural science is to transform the nonlinear problem into the linear problem and establish the linear system model on the premise of excluding the nonlinear factors. Under the influence of this scientific tradition, people think that the objective world is a set of objects characterized by simple linear relations. However, the complexity theory holds that there are very few linear dynamic systems that make definite and regular movements in the real world, and the world is nonlinear and complex essentially. Therefore, people must consider many factors to comprehensively and accurately understand the objective world. For a nonlinear complex system, the following points should be paid attention to:

(1) If you want to have a comprehensive understanding of the essence and state of a nonlinear system, you should ask questions from different aspects, at different levels, and in different ways.
(2) When a very small uncertainty in the nonlinear system is continuously amplified or reduced through the feedback process, a sudden change may occur at a bifurcation point, which eventually makes a simple system has surprisingly complex, thus making it impossible to make a long-term accurate prediction of the whole system. Most nonlinear systems existing in nature and human society are neither completely ordered nor completely random, which makes its behavior is between completely predictable and unpredictable.

From the perspective of ontology, nonlinear thinking holds that the real world is essentially nonlinear, but the degree and manifestation of nonlinearity vary greatly. A linear system is only an acceptable approximate description of a nonlinear system under simple circumstances. From the perspective of methodology, nonlinear thinking holds that nonlinearity should be treated as nonlinearity generally, and it can be simplified as linearity only in some simple cases. Nonlinear is thus the cause of the infinite diversity, unpredictability, and differentiation of the system and is a major source of complexity. Nonlinear thinking is a kind of thinking mode that directly faces the complexity of things themselves and the complexity of the mutual relations between things and understand the objects clearly beyond lineal thinking. As our thinking paradigm shifts from linear (atomism, reductionism) to nonlinear (systems theory), the understanding on nature and society is gradually profound.

1.2.2 *Uncertainty*

Since the 1960s, the research on chaos in modern system science has broken the ideological imprisonment of dividing "certainty" and "uncertainty" in traditional science. A large number of objective facts and experiments are used to indicate that colorful world is formed because of mutual connection and transformation between certainty and uncertainty. The research results on "uncertainty" in many disciplines have revealed the inevitable existence of uncertainty in the micro and macro world. For example, Heisenberg's uncertainty principle in quantum mechanics, Godel's theorem in mathematical logic, Arrow's impossibility theorem in social choice theory, and fuzzy logic all provide preparation conditions for "uncertainty" to become the object of scientific research from different subject aspects. Many scientific successes are because scientists have learned to use uncertainty in the pursuit of knowledge. Uncertainty is a driving force to promote scientific progress.

In essence, uncertainty originates from the inherent and internal hierarchy, openness, dynamics, coherence, nonlinearity, criticality, self-organization, self-reinforcement, and mutation of the social system itself. Based on the characteristics of uncertainty, uncertainty can generally be divided into five categories: objective uncertainty, subjective uncertainty, process uncertainty, game uncertainty, and mutational uncertainty.

1.2.3 *Self-organization*

Organization refers to the ordered structure in the system or the formation process of this ordered structure. German theoretical physicist Hacken divided "organization" into other-organization and self-organization according to its evolutionary form. Self-organizations are relative to other-organizations that cannot organize themselves, create themselves, evolve on their own, and cannot move autonomously from disorderly to orderly. Other-organization can rely only on specific instructions from the outside world to put the organization toward orderly evolution, thus passively moving from disorderly to orderly. On the contrary, self-organization refers to a system that can organize, create, and evolve itself without specific external instructions and can independently move from disorder to order and forming a structured system.

Self-organization theory is established and developed in the late 1960s. It mainly focus on the formation and development mechanism of complex self-organizing systems (life systems and social systems), to study how the system spontaneously moves from disorder to order and from low-level order to high-level order under certain conditions. It believes that the self-organization theory consists of the dissipative structure theory, synergetics, the mutation theory, the hypercycle theory, the fractal theory, and the chaos theory. Among them, the theory of dissipative structure is used to solve the condition environment problem of self-organization, synergetics is used to solve the dynamic problem of self-organization, the catastrophe theory is used to study the problem of self-organization from the perspective of mathematical abstraction, the hypercycle theory is used to solve the combination form problem of self-organization, and the fractal theory and the chaos theory are used to study the complexity and prospect problem of self-organization from the perspective of time sequence and spatial sequence. It is generally believed that the open system, far from equilibrium, and nonlinear interaction and fluctuation are the basic conditions for the formation of self-organization.

The phenomenon of self-organization is widespread both in nature and in human society. The stronger the self-organizing function of a system, the stronger its ability to maintain and generate new functions. We call this characteristic, which does not need external control and interference to achieve order through the adjustment and evolution of the system itself is self-organization. For example, Darwin's "the survival of the fittest" can be regarded as a self-organization process that organisms in nature achieve evolution through the self-adjustment of the ecosystem.

1.2.4 *Emergence*

Complexity science refers to the attributes, characteristics, behaviors, functions, and other characteristics that system as a whole has but parts or sum of parts haven't as emergence. In other words, when we reduce the whole to various parts, these attributes, characteristics, behaviors, and functions of the whole cannot be reflected in a single part. The emergence theory is one of the complexity theories put forward by Holland, a well-known scholar at Santa Fe Institute. He devotes himself to study different systems and models and reveal the common laws in the evolution process that elements make up the system from a low-level to a high-level. Emergence is commonly "the whole is greater than the sum of its parts". In his book *Emergence*, Holland demonstrated "simply creates complexity" by analyzing the footwork of checkers and the neural network. The core ideological content of emergence theory is the new quality with uncertain direction caused by interference and coupling of internal factors of the system under marginal conditions. Holland has used a metaphor to explain complexity through different phenomena, which provides a good way to explore complexity.

The assertion "born in nothing" by the ancient thinker of China, Laozi, is an ancient and profound understanding and expression of emergence. Bertalanffy took Aristotle's famous proposition that "the whole is greater than the sum of parts" to express emergence. Holland believes that the essence of emergence is "from small to large, from simple to complex". Complexity scientists often use "complexity comes from simplicity" to express emergence, they believe that complexity emerges from the simplicity in the evolution of things. Although emergence is a phenomenon and characteristic of the whole, the phenomenon and characteristic of the whole are not necessarily emergence. Bertalanffy distinguishes the two characteristics of whole, of accumulation and generation (non-addition) and divides the whole into two types: non-system sum and system sum. It should be recognized that the overall characteristics formed by adding up the characteristics of each part are not emergent, and only the generative (not additive) formed by the characteristics that depend on the specific relationship between parts can be called "emergent". It can be concluded from this that it is impossible to predict the emergence phenomenon from the simple addition of parts themselves. Emergence is a scientific concept that describes the model, structure, or characteristics presented by complex system levels.

1.3   Key Research Methods of Complexity Science

Complexity science is a new marginal and interdisciplinary subject, which has attracted more and more scholars' attention. Complexity science breaks the traditional paradigm of linearity, equilibrium, and simple reduction and devotes itself to studying various new problems brought by nonlinear, disequilibrium, and complex systems.

1.3.1   *Theoretical analysis*

Theoretical analysis is an essential way to study complex systems. It mainly includes theoretical analysis of complex systems before, during, and after the event. However, the judgment of complex systems is an important part of theoretical analysis in advance, and there are also some qualitative and quantitative judgment methods.

1.3.2   *Models*

Building models is a great creation of human beings in the practical process of understanding and transforming the world and is also the most commonly used method of scientific research. In its long history of making and using models, we have accumulated a wealth of experience and gradually developed universal model methods, which are ancient methods of scientific research and central to modern scientific methods (Sun and Zhang, 2004). The combination of both reductive and holistic properties, in particular the help of modern computer technology, making model method becomes a particularly important methodological tool in the study of complexity science.

In scientific research activities, the object entity is given the necessary simplification to portray its main features with appropriate manifestations or rules, so that the obtained imitation is called a model and the object entity is called a prototype. Among all living things on the earth, it is unique for human beings to establish certain entities or scripts to build models. The model is not a part of daily life, it mainly comes from the application of a certain work.

The important value of the model is that we can predict the results without having to carry out time-consuming, practices, especially for dangerous work. Even those scale models (steamboat models, aircraft models, railway models, etc.) will give us some quantitative data, otherwise, it would be difficult to get a real measure.

The widespread use of models plays a key role in the study of complexity. Complexity science generally establishes models of complex systems based on metaphorical analogy. To explore complexity, scientists have established a large number of complex system models from different aspects.

At present, there are four important models in complexity science:

(1) *Echo model of a complex adaptive system (CAS)*: When Holland studied CAS, he used a model method to construct the CAS model based on a metaphor, thus establishing his complex adaptive system theory. He established some rules by selecting building blocks and reorganizing these building blocks in different ways to create an easy-to-understand system model governed by certain rules, that is, stimulus–response model. This model reflects the basic behavior model of active subjects in CAS, that is, the understanding and description of how individuals adapt and learn. He established this micro-model in three steps: (1) establishing a model of the execution system; (2) establishing the credit distribution mechanism; (3) providing means of rule discovery. Based on the micro subject model, Holland began to establish a macro model of the whole system, which is called the echo model. The whole system includes several positions, each position has several subjects, and the subjects communicate with each other to exchange resources and information. This is the basic echo model.

This basic echo model is too simple to describe complex system behaviors, so Holland gradually introduced five mechanisms "exchange conditions", "resource conversion", "adhesion", "selective mating" and "conditional replication" based on the basic model to form an extended echo model. Through this step-by-step expansion, the expression and description ability of the echo model is continuously enhanced, thus having the ability to describe and study various complex systems. It is not hard to see that the echo model has a clear economic, biological, ecological context and impact. Therefore, the CAS theory is not a product of pure theoretical thinking, but a theoretical model that is gradually abstracted based on the study of a large number of real complex systems. The Santa Fe Institute also developed a software swarm to achieve the operability of the model. The echo model shows the complexity and emergence of the imitated system, but a large number of details are deleted. The echo model is an extremely beautiful model constructed with few principles, setting a signpost for how complexity emerges and adapts.

(2) *Generative model in emergence theory*: The emerging concept (that is, the whole is greater than the sum of its parts) is surprisingly simple, but it has profound implications in many fields such as science, commerce, and art. However, it is only a metaphorical description. To complete the expression by scientific language, various scientific models must be constructed. Under the guidance of complexity thought, Holland also continuously constructed various scientific emergence models based on a metaphor, thus revealing the complexity law hidden behind the emergence phenomenon. In his book *Emergence*, which deeply explores the phenomenon of emergence, Holland compares different systems and models that show the phenomenon of emergence and shows the common rules or laws between them.

Holland started with simple chess games, numbers, and building block models and then used the map metaphor and game theory to establish a dynamic model that reflects the invariant laws that lead to structural changes. With the aid of a computer, through the metaphorical analogy of checkers and the introduction of neural network theory, a subject-based emergence model with universal theoretical significance is established. Finally, through the analysis of the restricted generation process and checkers program and embedding genetic algorithm, Holland established a restricted generation process model with a variable structure. Holland vividly shows us through various emergence models that the emergence theory can predict many complex behaviors and at the same time give us many revelations about life, wisdom, and organization.

(3) *Sandpile model of self-organized criticality theory*: Self-organized criticality (SOC) was proposed in 1987 by three physicists Bak, Tang, and Wiesenfeld of Brookhaven National Laboratory in the

United States. It is used to describe and explain the complex behaviors such as $1/f$ noise spectrum (Bak *et al.*, 1987). The so-called self-organized criticality refers to a kind of complex systems which are open, dynamic, far from equilibrium and composed of multiple units that can evolve to a critical state through a long self-organized process. A small local disturbance in the critical state may be amplified by a mechanism similar to the "domino effect", and its effect may extend to the whole system, forming a large avalanche. Critical system is characterized by an "avalanche" of events of various sizes, and the sizes (time scale and space scale) of "avalanche" obey "exponential" distribution.

The concept of self-organized criticality is also metaphorical. To turn it into a scientific concept, a scientific model must be established. At first, scholars studied the coupled pendulum and later found that the situation of the coupled pendulum was not easy for people outside the physics to understand. They found that sand pile was a more vivid example, so they established a classical model of the self-organized criticality, that is sandpile model.

The data simulation of the sandpile model shows that the open, multi-degree-of-freedom, and far from equilibrium dynamic system in the critical state, evolves in the paroxysmal, chaotic, and avalanche-like form to a stable self-organized critical state. This model has been widely used in phenomena such as solar flares, volcanic eruptions, economics, biological evolution, turbulence, and the spread of infectious diseases. This simple sandpile model showed the secrets of the complex systems evolution.

There are many models in the study of self-organized criticalities, such as the Bak–Sneppen evolution model, Bak–Chen–Tang forest fire model, traffic congestion model, OFC (the Olami–Feder–Christensen) earthquake model, and so on. The model method occupies a very important position in the study of self-organized criticality. It is with the help of a model that we can understand the characteristics of some systems. Generally speaking, the self-organized criticality model is relatively simplified. Due to the complexity of the multi-body problem itself, if the model is too complicated, it will be difficult to deal with. Therefore, although most of the self-organized criticality models depict roughly, they can still reflect the basic framework, basic images, and the essential mechanism of the physical system and have a certain degree of universality.

(4) *Artificial life model in artificial life research*: Researchers also use model methods when studying the complex phenomenon of artificial life. For example, based on the metaphorical concept of chaos edge, Langton and other scholars have established various models to explore the generation and evolution of artificial life, such as self-propagating cellular automata, boids model, ant colony model, tierra model, avida model, and amoeba world. It is through these models that Langton *et al.* discovered that the essence of life lies in the organizational form of matter rather than in the specific matter itself, if we create conditions in a certain medium to produce chaotic edges, then we may create life in this medium.

1.3.3 *Numerical calculation and simulation*

1.3.3.1 *Numerical calculation*

Numerical value and science are inseparable, and the numerical method is already a relatively mature scientific method. There is also a specialized branch of numerical analysis in mathematics, and numerical methods are also specialized in scientific research and is referred to as "the third scientific method". It seems that discovering new phenomena by digital games and numerical values is fully developed in complexity science, so that numerical methods have become an indispensable method in complexity science.

Von Neumann had proposed that the evolution of organisms can be studied by computational methods. However, his idea was neglected for many years. It was until the 1960s to 1970s that John Holland put forward and perfected the genetic algorithm theory and had practical application (establishment of optimization model) in the 1980s that evolutionary computation has received extensive attention.

The relevant numerical calculation methods mainly include genetic algorithm, evolutionary computation method, cell to cell mapping method, etc.

A genetic algorithm is a search algorithm based on individual genotype representation and through operations such as crossover and mutation. When the genetic algorithm is applied to practical problems, the problem is first encoded. This encoding is called individual, and the set of individuals is called the population. Each individual represents a potential solution to the problem. Then, starting from a group of randomly generated initial solutions, they undergo operations such as crossover, mutation, replication, and selection. Crossover is an operation to exchange part of genetic information between different individual objects, mutation is an operation to change some genes of individual objects into other alleles (generally in probability mode), replication is an operation to transfer the genes from one generation to the next generation, and selection is an operation to eliminate individuals with poor adaptability to the environment. With the operation of the algorithm, good individuals are gradually retained and combined, thus producing new individuals. Since every generation has a recombination of genes, novel building modules are often produced in groups, so that the genetic algorithm will quickly produce building modules that have double or triple advantages. If the combination of these building modules produces greater advantages, then individuals with these excellent building modules will be popularized to the whole population faster than before. As a result, the genetic algorithm quickly points to the answers of the existing problems, even if it does not know where to look for the answers in advance. In the mid-1960s, Holland proved the basic theorem of genetic algorithm, which he called graphical theorem, that is, in reproduction, crossover, and mutation, almost all gene groups with extraordinary robustness can develop exponentially in species.

Traditional evolutionary computation is a randomized computational model generated with simulating the biological evolution process of "natural selection" and "natural heredity" (Schwefel, 1981). Its origins date back to the 1950s and can be found in influential works such as Friedberg (1958) and Bremermann (1962). However, in the next thirty years, this field is still quite unfamiliar to the many scientific workers. The main reasons are the lack of a powerful computer at that time and the defects of the early evolutionary computation itself. Until the 1970s, groundbreaking work from Holland (1962) and Rechenberg (1965) slowly changed this situation. Especially in the last decade, academic activities on the theme of evolutionary computation have increased year by year, and applications-oriented research in evolutionary computation has penetrated almost all walks of life (Dasgupta *et al.*, 2013). The evolutionary computation technology has been widely used in all respects because it benefited from the great improvement in computer performance and, more importantly, evolutionary computation not only has intelligent characteristics of self-organized rows, self-learning, self-optimization, but also has intrinsic parallelism, simplicity of principle, excellent globality, broad application, and so on.

With the deepening research in various disciplines, more and more practical problems that are difficult to solve by traditional methods have emerged. Compared with traditional mathematical methods, the evolutionary computation method has the following characteristics:

(1) The processing object of evolutionary computation can be the parameter itself or a specific code formed by some mapping. The coding form can be matrix, tree, graph, set, string, sequence, chain, table, etc. This feature makes evolutionary computation have a wide applications.

(2) Evolutionary computation adopts a group search strategy, while traditional methods mostly adopt a single-point search strategy, which makes evolutionary computation has a excellent global performance and reduces the risk of falling into local optimum. At the same time, it also makes evolutionary computation itself easy to implement in large scale parallel, which can give full play of high-performance computer systems.

(3) Evolutionary computation does not rely on the knowledge of search space and other auxiliary information. It uses fitness function to evaluate individuals and drives the evolution process on this, without special strict requirements on fitness function itself. However, traditional methods require functions to have conditions such as continuity, differentiability, or spatial convexity. This gives evolution computing a wide applications.

(4) Evolutionary computation uses the transition rule of probability to control the search direction. It abides by some random probability and approaches the optimal solution in the sense of probability. Therefore, unlike the traditional method that usually uses deterministic rules to construct an appropriate descending direction.

The basic idea of the cell mapping method is to discretize the state space of a dynamical system into many small geometric bodies (cells) and the set of all cells forms cell space (Hsu, 2013). After the state space of the dynamic system is transformed into the cell space, the state transition in the dynamic system naturally corresponds to the transition between cells. Furthermore, the corresponding research on the dynamic system is completed through the research on the transfer relationship between cells. In the early 1980s, for its fast, accurate, and wide application, cell mapping method has attracted the attention of many researchers and has become a research hotspot since it appeared (Hsu, 1980).

Compared with the direct numerical simulation method, the cell mapping method takes cells as research units and forms the transfer relationship between cells (cell to cell mapping dynamic system) according to the transfer relationship of a dynamic system. After studying the cell mapping, the classification of cell space is obtained. Dynamic interpretation of classification results was carried out to complete the research on the dynamic system. The cell mapping method does not study the evolution of each trajectory in the dynamic system in isolation, but pays attention to the information transmission between each cell as a whole, avoiding processing the information isolated as the direct numerical simulation method, thus greatly improving the efficiency. Also, some dynamic behaviors that are difficult to be obtained by direct numerical simulation method can be obtained by in-depth analysis of transition information between cells (such as unstable invariant sets obtained by graph cell mapping method) (Hong and Xu, 1999, 2001). Moreover, based on the concept of the cell, the direct numerical simulation method can be improved to obtain a more efficient point mapping method (Tongue and Gu, 1988).

1.3.3.2   *Simulation method*

The main simulation methods are system dynamics method, cellular automata method and the swarm method.

The so-called system dynamics is a thinking way about the essence of the overall operation, integrating structural methods, functional methods, and historical methods into a whole. Its purpose is to enhance the "group intelligence" of human organizations. System dynamics is developed based on summarizing operations research to meet the management needs of modern social systems. It is not based on abstract assumptions, but on the premise of the existence of the real world, do not pursue the "best solution", but from the whole to seek opportunities and ways to improve the system behavior. Technically speaking, it does not obtain the answer according to the deduction of mathematical logic, but establishes a dynamic simulation model according to the actual observation information of the system and obtains the description of the future behavior of the system through computer experiments. Simply put, "system dynamics is a computer simulation method for studying the dynamic behavior of social systems". Specifically, system dynamics includes the following points:

(1) System dynamics analyses both living and non-living systems as information feedback systems and holds that there are information feedback mechanisms in each system, which is exactly the important viewpoint of cybernetics. So, system dynamics is based on cybernetics.
(2) The system dynamics divides the research object into several subsystems and establishes the causal relationship network among each subsystem. It focuses on the relationship between the whole and the whole instead of the traditional element view.
(3) The research method for system dynamics is to carry out a computer simulation test by establishing a computer simulation model, then the basis for the formulation of strategy and decision would be provided.

The cellular automata model was first created by John von Neumann in 1950. Since 1986, Wolfram has developed it to simulate complex phenomena. We know that there are many models for simulating physical systems. The time and state or function of the ordinary differential equation model are continuous. The time of the mapping model or difference equation is discrete, but the state is continuous. The cellular automata model is another simulation method of physical systems, in which time, space, and state function are discrete. The characteristic of the cellular automata method is to obtain the global spatial–temporal structure through the local dynamic evolution rules. The cellular automata method first divides the space into several cells, then the state of each cell is expressed by discrete quantities, and finally, according to physical considerations, the rules of how the cell evolve with time according to the states of neighbors would be set. According to the evolution of the above steps, cellular automata will form various complicated patterns after a long time. Based on a large number of computer experiments, the dynamic behaviors of all cellular automata can be summarized into four categories:

(1) *Smooth type*: Since any initial state, the cell space tends to a space-smooth formation running for a certain time, where space-smooth refers to the cell is in a fixed state with time.

(2) *Cyclical type*: After a certain time running, cell space tends to be a set of simple fixed structures (stable patterns) or periodical patterns. Since these structures can be regarded as a kind of filter, they can be applied to image processing.

(3) *Chaotic type*: After running for a certain period from any initial state, cellular automata shows a chaotic aperiodic behavior, and the statistical characteristics of the generated structure no longer change, usually showing fractal dimension features.

(4) *Complex type*: In this type, local complex structures (or chaos) appear, and some of them could propagate.

To test the feasibility of the echo model, the Santa Fe Institute developed a software platform swarm to realize the theory of complex adaptive systems. On this platform, most of the functions of the echo model have been realized, and swarm has verified the feasibility and correctness of the complex adaptive system constructed by Holland. The simulation idea of swarm is to establish a series of independent individuals and investigate the behavior and evolution law of the system through interaction between independent events. The basic simulation unit of swarm is an individual. An individual is like an actor in a system, which can produce actions to affect itself and other individuals. The simulation includes several groups of interacting individuals. For example, a simulation of an ecosystem can include wolves, rabbits, and carrots. In an economic simulation, individuals are companies, securities agents, dividend distributors, and central banks. Independent and interactive simulation between individuals is different from the continuous simulation that the simulation is often a quantitative relationship between a set of related equations.

Individuals define the basic objects in the swarm system, while timetables define the process of events occurring between these objects. In swarm, specific behaviors occur at specific times, and the development of behaviors is carried out according to the schedule. A timetable is a data structure that contains the execution order of various events. For example, in the wolf-rabbit simulation system, there may be three kinds of behaviors: "rabbits eat turnips", "rabbits avoid wolf's tracking", and "wolves eat rabbits". Each action is independent. In the timetable, the three behaviors are sorted in the following order: "every day, rabbits eat carrots first, then they avoid being tracked by wolves, and then wolves try to eat rabbits". The model will move forward according to the execution order of this scheduled event.

1.4   Complexity Science and Earth System Science

1.4.1   *Research theory and knowledge basis of earth system science*

1.4.1.1   *Concepts*

Earth system science studies the overall structure, characteristics, functions, and behaviors of the earth system and is the basic theory of global change. Its scientific goal is to clarify the law and mechanism of the interaction between natural and manmade driving forces and the changes of the earth system, reveal the law and mechanism of the overall evolution of the earth system, establish the changing trend of the earth system, and establish the regulation theories and methods of the earth system change. The realization of this goal has great scientific significance and application value for guiding the sustainable development of human society.

The term "Earth Science System" was first proposed in written form in the book "Earth System Science" published by the Earth System Science Committee established by NASA in 1988, but no clear definition was given. In the book "Our changing planet: an introduction to earth system science and global environmental change" published by Mackenzie and Mackenzie (1995), earth system science is listed as a separate item and given its definition, it is all knowledge that studies the earth as a whole, it focuses on various functions and interaction of atmosphere, hydrosphere, biosphere, and lithosphere.

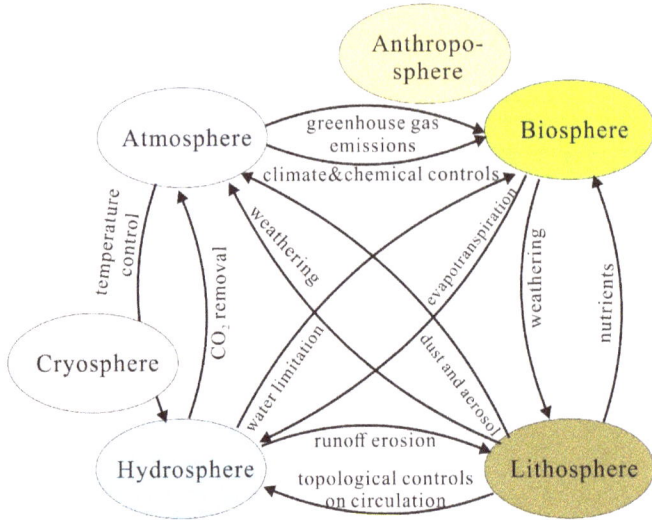

Figure 1.2. Relationship between different layers in the earth system (Cornell *et al.*, 2012).

1.4.1.2   *Complexity of earth system science*

In the layers of the earth system or the subsystems, (see Figure 1.2), the interaction between the components and the motion of the earth control many processes, for example, the rotation of the earth and the differential rotation of the earth's core, mantle, and crust, the original heterogeneity of the earth and the formation of super-large deposits, the formation of orogenic belts and basins, erosion and sedimentation, mineralization, crustal deformation, climate change, ocean circulation, the evolution of ecosystems, etc. The factors and functions that control a basic process are various and different under different circumstances. In the history of the earth, the earth's surface environment has always been in constant change, and there are gradual changes and abrupt changes such as the extinction and recovery of organisms, climate fluctuations, mutation in the ecosystem, and the outbreak of large-scale mineralization.

The research of system science shows that the earth is a very typical open complex dynamic giant system. It includes numerous, relatively independent, interactive, and interdependent systems of different levels, types, and functions. It has a multi-level structure, in which layers or subsystems, and various functions, all have different spatial and temporal scales from microscopic to macroscopic, and the processes of various scales are also interrelated (Ehlers and Krafft, 2006). This is consistent with two important conclusions drawn from contemporary earth science research, first, changes at planetary scales are the result of interaction among the earth's subsystems; second, any time scale change includes the interaction between the earth system on various time scales. Processes with shorter time scales will also affect the evolution of the earth, and geological processes are not necessarily long-term processes, while the long-term processes will affect short-term operational processes, and local processes will also affect global processes. For example, environmental changes on a global scale are inconsistent at different time scales or spatial scales. Therefore, the spatial–temporal structure of the earth system is complex.

As an open complex giant system, the earth system has seven basic properties (Yu, 1998):

(1) *Multi-component coupling and interaction*: The earth system is a multi-component coupling dynamical system with a diverse composition of the lithosphere, hydrosphere, atmosphere, and biosphere, and the complex interaction among them.
(2) *Openness and dissipation*: The earth system is open, it exchanges matter and energy with the environment through dissipation, and maintains its equilibrium steady-state and dynamic evolution.
(3) *Non-equilibrium*: A region usually undergoes a complex history of sedimentation, magmatism, tectonic movement, metamorphism, mineralization, weathering, denudation, and other processes, so as a whole, the geological body deviates from the equilibrium state.
(4) *The irreversibility of the processes and their multiple coupling*: The processes are irreversible and interactive for a long time such as the accretion of continental crust, the evolution of orogenic belts and basins, and the formation of metal deposits.
(5) *Nonlinearity*: The earth system is full of various nonlinear factors, such as nonlinear geochemical reaction, nonlinear water–rock interaction, nonlinear mineralization, and nonlinear land–sea interaction.

(6) *Random fluctuation*: There is a large degree of freedom in the earth system, resulting in fluctuation, such as the free oscillation of the earth.

(7) *The duality of continuity and discreteness of the earth's material structure*: The material in any geological region can be regarded as a continuous medium and constitutes a geological field. However, the material is composed of discontinuous molecules, atoms, and other particles, and the macro properties of the continuous medium and field depend on the properties and behaviors of microscopic particles.

These basic attributes determine the complexity and nonlinearity of the earth system. It can be said that earth as a complex system that has undergone a long evolution in the universe, its complexity and nonlinearity are ubiquitous. All the functions of the earth system have nonlinear characteristics and show complexity such as multi-scale, self-organization, order, and disorder.

1.4.1.3 *Research theory of earth system science*

Although the research of earth system science is in the ascendant, as a new discipline, it is not yet mature. Theoretically speaking, the earth system science studies the mechanism of the interrelation and interaction among the subsystems that make up the earth system, the laws of the earth system changes, and the mechanism of controlling these changes, to establish a scientific basis for the prediction of global environmental changes and provide a basis for the scientific management of the earth system. Further, the study of earth system science must use modern means such as observation, exploration, and information technology. The processes of solid earth system, fluid earth system, and biological earth system at different time and space scales and the interaction mechanism among the systems should be understood and simulated, and the mutual coupling and changes of the earth surface system composed of lithosphere surface, atmosphere, hydrosphere, and biosphere, etc., should be studied (Cornell *et al.*, 2012) to reveal the interaction mechanism between human activities and resources, environment and ecosystem. The space scope of earth system science research ranges from the center to the outer space of the earth, and the period ranges from decades to millions of years, even billions of years (Schellnhuber and Wenzel, 2012). Earth system science attempts to construct a unified system theory framework, focusing on the interaction and methods between the earth's structure and structural components to explain the earth's dynamic characteristics, the earth's evolutionary history, and global environmental changes, emphasizing new concepts such as "source, flow, sink, field and response".

Specifically, the research contents of earth system science include (Zhou, 2004):

(1) *Interaction of the outer drive and the earth system*: Celestial and human activities are important external driving forces for the changes in the earth system. They affect the overall structure, function, and behavior of the earth system on different spatial–temporal scales. Studying the dynamic, physical, chemical, and biological processes of this influence is an important content to reveal the laws and mechanisms of the changes of the earth system.

(2) *Layer interface dynamics*: The nonlinear coupling of each layer of the earth system is an important feature of the earth system change. This interaction is mainly manifested in the exchange of energy, momentum, and matter on the interface of each layer. This exchange is determined by the physical, chemical, and biological characteristics of the media on both sides of the interface and has different spatial–temporal scale structures. They have non-uniform structures with different scales in space and non-stationary processes with different scales in time. Therefore, it is an important content of earth system science to determine the interface flux function taking space, time, and the state of each layer as independent variables and to establish a unified interface flux dynamics model.

(3) *Evolution law and mechanism of the earth system*: The earth system is a complex giant system, and its evolution process has all the characteristics of nonlinear dynamics such as self-organization, mutation, chaos, and so on. Based on the comprehensive analysis of massive data on paleontology, paleoanthropology, paleogeology, paleomagnetism, paleoenvironment, paleoocean, paleoclimate, and astronomy obtained by different experimental methods in different time and space scales, the

theory and analysis methods of complex systems are established
to reveal the overall characteristics, laws, and mechanisms of the
formation and evolution of the earth system. In recent hundreds
of years, the influence of human activities on the various layers
of the earth system has intensified day by day. It is necessary
to study the overall influence of human driving forces from the
perspective of the earth system.

(4) *Earth system dynamics mode*: The earth system dynamics mode
is an important means to deeply reveal the law and mechanism
of system change and predict its change trend, and it is also an
important symbol for earth system science to move toward quan-
tification. The earth system dynamics mode can be established
from two aspects: on the one hand, it is integrated into the earth
system dynamics mode based on each layer dynamics equation
and the layer interface dynamics equation; on the other hand, the
nonlinear statistical method is used to construct the statistical
dynamics model of the earth system based on synthesizing the
massive spatial and temporal distribution data of the historical
evolution of the earth system.

(5) *Forecast of the future change trend of the earth system*: Forecast
of future change trend is an important application goal of earth
system science research, and it is also a basic scientific problem of
earth system science. The earth system is an open nonlinear complex giant system, and its variation is a non-stationary process.
The predictability and prediction method of the mutual transformation of deterministic process and stochastic process for such
systems are still unsolved cutting-edge scientific theoretical problems. Given the certainty and randomness of the earth system
process at different spatial–temporal scales, the main content of
the research is to establish new prediction theories and methods
by applying the earth system stochastic dynamics model.

(6) *Regulation of earth system changes*: The energy of the earth system change process is very huge. Apart from the energy of celestial bodies acting on the earth, the energy of human activities is
still a small amount. However, a nonlinear complex giant system
is often in an unstable state, and its evolution is often a process
of continuous transformation from stable to unstable. As a small
disturbance, human activities may regulate the development process of the earth system when it is in an unstable state. The

Antarctic ozone hole, greenhouse effect, and artificial precipitation enhancement are all typical examples. This fully shows that studying the sensitivity of the earth system changes to disturbances and establishing the theories and methods of the earth system regulation can design human social and economic activities coordinated with the earth system changes and establish the earth management system to ensure the rapid, healthy, and safe sustainable development of society.

1.4.1.4 *Cognitive basis of earth science system*

The research thinking of earth system science is very clear (Wang *et al.*, 2003). It is using modern thinking mode and scientific systematic perspective to regard the earth system as an extremely complex self-organizing system with hierarchical structure and continuous evolution in the social environment system. The research steps of earth system science are: firstly, observation and accumulation of phenomena; secondly, starting from the laws of physics, chemistry, and biology, the quantitative relationship of the earth's process is established and then explained; thirdly, establishing a model; fourthly, verifying the model.

The research of earth system science must have the following supporting conditions:

(1) *Earth system integration database*: Geoscience has accumulated a large amount of quantitative historical evolution data of each layer of the earth system. These data have different spatial and temporal scale, different acquisition methods, and different measurement accuracy. These data can be used in the scientific research of the earth system after assimilating and integrating under unified rules. Therefore, the establishment of the earth system integration database is necessary to conduct earth system science research.

(2) *Earth system exploration network and integrated scientific experiments*: Taking nature as the laboratory to carry out scientific observation experiments, systematically and continuously providing unified observation experimental data is the basic support for earth science research. What the earth system science requires is not only the data observed separately of each layer of the earth system but also the comprehensive data observed of

the earth system. In the future, the earth environment satellite exploration, combined with the earth system exploration network of ground observation and comprehensive scientific experiments organized by special topics, will be the technical support for the digitization of the earth system and the important foundation for the research of the earth system science.

(3) *Earth system numerical simulation laboratory*: The earth is a multidimensional system, that the evolution of its state is a vast amount of spatial–temporal distribution data as well as multidimensional nonlinear dynamic equations. Using giant high-speed computer technology for nonlinear statistical analysis and numerical simulation is the only effective method to analyze the data. The earth system numerical simulation laboratory, combined with information provided by the earth system integration database and the earth system observation network, support the development of scientific research in the earth system.

1.4.2   *Needs and history of the transition of traditional earth science to earth system science*

The development of science and technology, especially the profound changes in the field and mode of global competition, has made the transformation of traditional geological work to modern geological work focusing on earth system science an irresistible trend. With the deepening understanding of earth system science, the overall concept of the earth system science has become the focus of experts in many disciplines. It has become an important topic for education, science and technology, and management departments related to earth science to deeply discuss the laws of earth system science from all-round and multi-perspective (Lohmann *et al.*, 2015).

Traditional earth science focuses on the in-depth study of a certain field, and this research method is suitable for the development stage of human social practice. It has made many achievements in many fields, forming complete discipline systems as well as traditional research methods and knowledge systems with their characteristics, such as meteorology, oceanography, geography, geology, and ecology (Lohmann *et al.*, 2013). Due to the wide space, the long history of formation, and the complexity of constituent elements, it still cannot fully understand the earth, although some disciplines have reached quantitative and semi-quantitative level. On the other hand, with the further deepening of human social practice, traditional geoscience is facing severe challenges. These challenges come more from the constraints of human society's development: the use of fossil fuels and agricultural reclamation release a large amount of CO_2, plant available nitrogen coming from industrialized agriculture and the use of fossil fuels is much higher compared with natural processes, and humans have used about 50% of the fresh water, animal and plant extinction rate is also much higher driven by human activity compared with its natural extinction (Turner, 2018). This man-made influence on the environment makes the earth system develop in a direction that is not conducive to the progress of human society (Barnosky *et al.*, 2014).

Till the 1980s did scientists generally realize that the earth must be studied as a unified system composed of various interacting components or subsystems, mainly the core, mantle, soil and lithosphere, atmosphere, hydrosphere, and biosphere (including human society). Only in this way can we truly deepen the study of the earth, and answer a series of urgent environmental problems faced by human beings in the earth system. This change of horizons and concept marks the traditional earth science transfer to earth system science. The realization of this transformation has a dual background. On the one hand, it is the necessity of the in-depth development of various branches of earth science. Many interdisciplinary subjects have emerged in traditional earth science, such as biogeochemistry, global change, and climate systematics. Moreover, according to the representative research in various fields of earth science at present, systematization, cross-cutting, and marginalization are the common characteristics of these results. On the other hand, the rapid development of space technology in recent decades has broadened human horizons and greatly improved humans' ability to understand the earth.

Under the dual needs of human social and economic development and the development of earth science, earth system science emerged. In NASA's book *Earth System Science* in 1987, the concept of "Earth System Science" were put forward firstly. Under the guidance of the earth system science, a series of important international joint research programs have been organized on a global scale in the 1990s and early this century to carry out comprehensive observation on the physical, chemical, and biological processes of the earth system elements and layers on a regional, national, and even global scale (Ehlers and Krafft, 2006). Some profound programs are the Global Environmental Monitoring System (GEMS), the Global Terrestrial Observing System (GTOS), the Global Climate Observing System (GCOS), the Global Ocean Observing System (GOOS), the International Long-Term Ecological Research (ILTER), and the Global Flux Network (FULXNET). These networks have laid a good foundation for the development of earth observation systems on a global scale.

The Earth System Science Partnership (ESSP) was established in Amsterdam in 2001. It is composed of the International Geosphere-Biosphere Program (IG-BP), the World Climate Research Program (WCRP), and the International Human Factors Program for Global Environmental Change (IHDP). At present, ESSP has initially formed a system with multiple functions including scientific research, analysis and simulation, capacity building and cooperation, specifically including four tasks, joint program, earth system analysis and simulation, regional integration research and capacity development, scientific liaison and academic exchange.

Till 2015, earth system science has been relatively mature and begins to carry out major institutions restructuring and interdisciplinary research on high-level. IGBP (International Geosphere-Biosphere Program), IHDP, and DIVERSITAS were combined into the new plan "Future Earth" in 2015, aiming to accelerate the transformation to global sustainability research through innovation. The plan is based on the research of the early global change has worked more closely with management departments and private enterprises to jointly design and study new knowledge for a sustainable future.

Although earth system science has made great progress, it still faces great difficulties. First of all, in the face of the complex and open giant system of the earth system, how can we timely and periodically obtain the massive data of multiple parameters of the system? How can the model be checked in due course? How can we share and exchange scientific research and valuable data obtained by thousands of geoscience laboratories, scientific research institutions, colleges, and universities all over the world? The most important thing is, in the face of so many subsystems, how should earth system science gradually integrate some of them to realize the optimal system integration and expansion. This is also the evolution path of this discipline, which is bound to be limited and supported by modern engineering and technology science. Another difficulty may lie in how to integrate ecosystems with natural systems. Because the research of natural systems is about natural laws, those are not subject to people's will (although natural subsystems may show nonlinear relations, resulting in complex phenomena, it can be predicted that with the development of complex science, this difficulty can be overcome); however, social science research (ecosystem) is about social laws, which are determined by biological behaviors. The accuracy (existence) of social laws has always been the focus of debate among social scientists. When the earth system science tries to perfectly integrate the ecosystem and the natural system, it will inevitably involve the psychological research that determines the biological behavior, which would be a major difficulty in establishing the model.

To sum up, the development of contemporary technology, especially space technology and large computers, has made the overall exploration of earth a reality (Schellnhuber, 1999); the maturity of many branches of earth science has promoted their interconnection and interdependence. Contemporary earth science is changing from taking branches of disciplines as the main body to a new era of cross-penetration among disciplines to study the earth system. The earth system science is to study the complex interaction process between various components and layers of the earth system in a very wide range from the earth's core to the earth's outer space, and the mechanism of controlling these processes, it is required that earth science must be combined with life science, chemistry, physics, mathematics, information science, and social science to understand the evolutionary history of the earth and establish a scientific basis for predicting global environmental changes to improve the living environment of human beings and maintain sustainable social and economic development.

Earth system science cannot replace the development of traditional earth science disciplines themselves. On the contrary, it requires them to study and provide more in-depth and accurate knowledge of each element of the earth system. From the aspects of research objects, research methods, and problems to be solved, the earth system science has many brand-new characteristics and higher levels than the traditional earth science and is one of the most valued emerging sciences in the late 20th and the 21st century.

Chapter 2

Complexity of the Groundwater System and Its Influencing Factors

2.1 Groundwater System — Open and Complex Giant System

Groundwater is an important source of fresh water in the world. More than 1.5 billion people in the world rely on groundwater as their main source of drinking water (Alley *et al.*, 2002). Groundwater is an important part of the hydrological cycle and is very important to maintain rivers, lakes, humidity, and aquatic biological communities.

Groundwater systems include subsurface water, geological media in which water is stored, flow boundaries, sources (e.g., surface water recharge), and sinks (e.g., springs, interlayer flows, or wells). It consists of water-bearing system and groundwater flow system. A water-bearing system generally refers to the water-bearing medium trapped by water-resisting or relatively water-resisting stratum, and they have a unified hydraulic connection (Wang *et al.*, 1995); a groundwater flow system refers to a groundwater body composed of flow surface groups from source to sink with a unified temporal and spatial evolution process (Eengelen *et al.*, 1986). Under natural conditions, the flow time of groundwater from the recharge area to the discharge area ranges from less than one day to more than one million years. The age range of water present in the system is from the recent precipitation to water formed by sedimentation in geological history (Alley *et al.*, 2002). Therefore, there is a material exchange between the groundwater system and the outside world, it is an "open and complex" giant system. When studying the groundwater system, its "complexity" should be fully considered, instead of starting with its simple definition.

2.1.1 *Complexity of runoff pattern*

The definition of groundwater system reflects the characteristics of integrity. The groundwater system is a typical "complex" system, and its research content is very rich, including the runoff flow pattern, the runoff process, the system boundary, and the system time. In traditional hydrogeology, laminar flow and turbulent flow are divided according to the Reynolds number (the ratio of inertial force to viscous force when fluid flows). When the Reynolds number is less than 500, laminar flow is considered, and when the Reynolds number is greater than 2000–4000, turbulent flow is considered. The applicable condition of Darcy's law (Q = KAI, Where, Q is the rate of water flow (mL/s), K is the hydraulic conductivity (cm/s), I is the hydraulic gradient (dimensionless), and A is the column cross-section area (cm^2)) obtained by scholars based on laminar flow experiments is laminar flow with Reynolds number less than 10, and hydraulic gradient I is used as the driving force of water flow in a homogeneous isotropic medium field, which is only a special case when the permeability coefficient K is constant (Zhang *et al.*, 2011). Based on Darcy's law, a series of concepts and parameters of groundwater dynamics were derived. These theories and concepts will no longer be applied to laminar with Reynolds numbers greater than 10 or turbulent flows or "free convection" (Wood and Hewett, 1982).

In contemporary studies, groundwater with high velocity is often treated differently in combination with its geological occurrence. However, considering the complex of groundwater runoff, the driving force of non-laminar water flow should be characterized by infiltration velocity V (Zhang *et al.*, 2011). Velocity is positively correlated with Reynolds number, and when the velocity exceeds a certain critical point, underground runoff shows a turbulent flow characteristic, the flow line swings in waves, and the frequency and amplitude of the swing increase with the velocity. As for particles in the flow field, their flow velocity and direction change irregularly with time, so various physical parameters of fluid such as velocity, pressure, and temperature, etc., all change randomly with time and space. Turbulence is a nonlinear and multi-scale fluid flow pattern with irregular space and unordered time. "Free convection" is a runoff movement driven by buoyancy, that is formed by density difference. The density difference can be caused by temperature difference or salinity difference, which is often ignored in groundwater dynamics. Therefore, it is universal to fully consider the complexity of groundwater runoff flow patterns and use infiltration velocity V to characterize the groundwater driving force of complex systems, which provides new explanations for many hydrogeological phenomena.

2.1.2   *Complexity of the runoff process*

Groundwater is affected by the hydrodynamic field, the hydrochemical field, the temperature field, the pressure field, and human activities in the runoff process, resulting in complex water–rock interactions, including dissolution/precipitation reaction, adsorption/desorption reaction, redox reaction, and microbial action. Complex stratigraphic media combined with complex biogeochemical reactions eventually form diversified groundwater change processes and patterns.

The hydrodynamic field controls the path and flow pattern of groundwater runoff, including the evolution process of recharge, runoff, and discharge, thus forming groundwater flow systems of different scales (local flow system, intermediate flow system, and regional flow system) (Figure 2.1). In addition to the runoff process of groundwater itself, the dynamic field also affects the material exchange between the groundwater system and the outside world, such as groundwater–surface water interaction and groundwater–soil water interaction (Du *et al.*, 2018). The hydrodynamic field is an indirect embodiment of potential energy. The research content involves the seepage process in porous media, related laws and equations, definite conditions, and the establishment of mathematical models. "Groundwater Dynamics" specializes in studying the movement law of groundwater in pores, fissures, and karsts, including the movement of groundwater to canals (divided into the stable movement and unstable movement); the law of groundwater movement in irrigation and drainage areas; theory of groundwater movement in the unsaturated zone (infiltration, unconfined water evaporation, etc.); hydrodynamic dispersion theory; and solute transport law

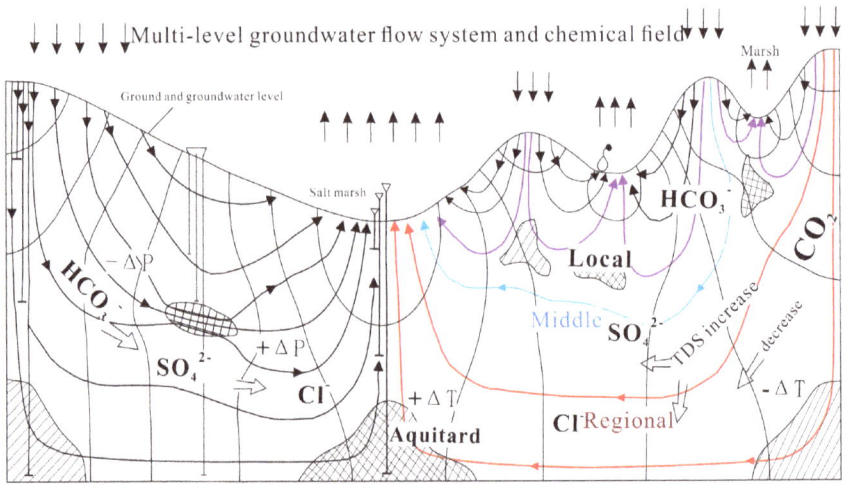

Figure 2.1. Groundwater flow network and system structure.

in groundwater, water transport law in the vadose zone, etc. The complex hydrodynamic field has a profound influence on the runoff process of groundwater, not only controlling the flow, but also exerting a certain influence on the hydrochemical field and thermal field. For example, the process of recharge and discharge will cause solute migration, forming the "complexity" in the runoff process.

From the point of hydrogeochemistry, the key elements and minerals in the groundwater system have a series of reactions in the process of groundwater runoff, such as C, N, S, Fe, and microorganisms participating in redox reactions, which have a profound impact on the hydrochemical characteristics of groundwater, but the related processes are very complicated. Organic carbon, as the carbon source and electron donor of microorganisms, affects a series of redox reactions (Figure 2.2), and its mineralization decomposition will lead to the release of its adsorbed harmful components (such as organic pollutants, heavy metals, or ammonium nitrogen) (Gatland *et al.*, 2014). The redox process of nitrogen is relatively active, including nitrification/denitrification, ammoniation, and anaerobic ammonia oxidation. Nitrogen circulation in the groundwater system is also an important factor to regulate the chemical composition and characteristics of groundwater (Du *et al.*, 2017). The geochemical cycles of sulfur and iron are also important and complex; sulfur mainly takes

Eh(mV)	Redox point pair	Reaction	
800	O_2/H_2O	$CH_2O + O^2 \rightarrow CO_2 + H_2O$	Respiration
600	NO_3^-/N_2	$CH_2O + \frac{4}{5}NO_3^- + \frac{4}{5}H^+ \rightarrow \frac{2}{5}N_2 + CO_2 + \frac{7}{5}H_2O$	Denitrification
	MnO_2/Mn^{2+}	$CH_2O + 2MnO_2 + 4H^+ \rightarrow CO_2 + 2Mn^{2+} + 3H_2O$	Mn reduction
400	NO_3^-	$CH_2O + \frac{1}{2}NO_3^- + H^+ \rightarrow CO_2 + \frac{1}{2}NH_4^+ + \frac{1}{2}H_2O$	Ammonification
200	$Fe(OH)_3/Fe^{2+}$	$CH_2O + 4Fe(OH)_3 + 8H^+ \rightarrow CO_2 + 4Fe^{3+} + 11H_2O$	Fe reduction
0			
	SO_4^{2-}/H_2S	$CH_2O + \frac{1}{2}SO_4^{2-} \rightarrow \frac{1}{2}H_2S + HCO_3$	Sulfate reduction
-200	CO_2/CH_4	$2CH_2O \rightarrow CO_2 + CH_4$	Methanogenesis

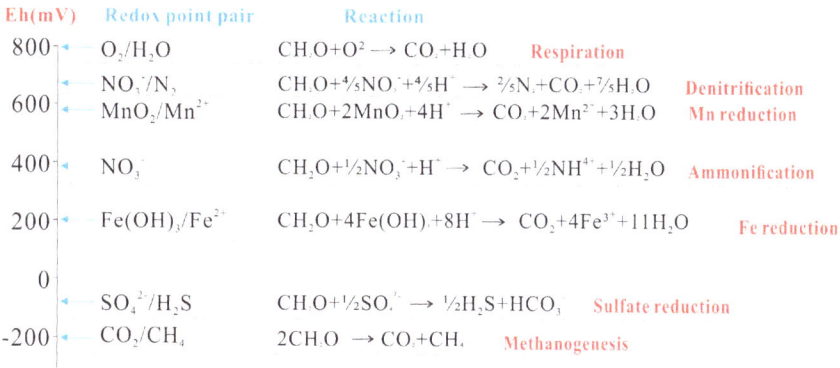

Figure 2.2. Redox reaction sequence involving DOM in natural groundwater (Liu *et al.*, 2020b).

part in desulfurization acid action and participates in precipitation reactions related to iron, such as FeS precipitation, while the iron is an element with relatively active redox properties in groundwater system, which controls the adsorption and release of heavy metals, mineralization and decomposition of organic matter, and interaction with other key elements (Liu *et al.*, 2020a). The above elements are only typical elements that control groundwater chemical changes, and other elements such as halogen play a role in the runoff process of groundwater. Microbial action is the engine of a biogeochemical reaction. Microorganisms with different community structures play different mediating roles, such as sulfate-reducing bacteria and iron-reducing bacteria (Pan *et al.*, 2016) with different compositions are also the main factors that affect the chemical composition of water. Water–rock interaction causes the composition in a solid medium to dissolve/precipitate. For example, flowing through strata with a high gypsum content leads to an increase in sulfate content in groundwater, and flowing through strata with limestone leads to an increase in carbonate and bicarbonate content in groundwater. Compared with the dynamic field, the chemical field of the groundwater system is also a "complex system" affected by multiple processes, which mainly affects the hydrochemical composition of groundwater through a series of reactions during groundwater runoff.

Temperature (thermodynamic) and pressure fields also constitute the "complexity" of groundwater systems. For example, shallow aquifers are affected by seasonal temperature changes, while deep confined aquifers may be affected by uneven geothermal distribution (Bravo *et al.*, 2002). When geothermal water containing calcium bicarbonate is exposed to the surface, a large amount of carbon dioxide escapes due to the reduction of pressure to form a chemical precipitate of calcium carbonate. With the development of industry, agriculture, intense human activities mainly affect the runoff process of groundwater by changing the runoff conditions and releasing pollutants. The above factors are the key factors that affect the hydrochemical characteristics of groundwater, and this is only the tip of the iceberg of the "complexity" for the groundwater system. A complex medium structure or system boundary is also an easily neglected part in the research.

2.1.3 *Complexity of system boundaries*

The system boundary includes the vertical bottom boundary of the system, the boundary reflecting the aquifer structure and the boundary including ecological subsystems. The fuzziness and uncertainty of these three types of boundaries lead to complexity. The fuzziness of the vertical bottom boundary can be caused by the difference of seasonal rainfall, which leads to the change of runoff depth, followed by the change of land subsidence compaction release during groundwater collection (Konikow and Neuzil, 2007), which makes the real geometric boundary of aquifer system fuzzy and unstable.

In the field of hydrogeology, stratum are divided into aquifer, aquitard and aquiclude, and it is believed that low permeable medium such as aquitard and aquiclude can only transmit a certain amount of water vertically, but cannot provide a certain amount of water by themselves. However, the impact of the compaction process of low permeability medium on groundwater quality and quantity cannot be ignored and has gradually attracted worldwide attention (Hendry and Woodbury, 2007; Waber and Smellie, 2008; Jiao *et al.*, 2010). In the case of Quaternary sediments, the permeability and connectivity of loose sediments are gradually reduced under the compaction driven by gravity, and complex water–rock interactions occur within them, then the sediments gradually evolve into clay (Potter *et al.*, 2005). As a common low permeability medium in plain areas, the clay layer is regarded as a typical weak permeability layer in the groundwater system (Du *et al.*, 2017). Compaction prolongs the time of water–rock interaction by reducing permeability. Also, it releases a large amount of pore water originally occurring in sediments into adjacent aquifers due to static load pressure exceeding the sum of effective stress and pore fluid pressure, thus affecting the quality and quantity of groundwater (Wang *et al.*, 2013). Toxic pollutants imported from human industrial and agricultural activities can participate in the hydrological cycle along with the runoff of groundwater, but the harmful components of "natural sources" in groundwater cannot be ignored, either. The research on the contribution of low permeability media to aquifer has become a hot topic in the research of groundwater pollution sources. In addition to several studies have found that a large part of the groundwater comes from the compacted release from the aquitard (Beylich *et al.*, 2010), for the migration and transformation process of pollutants, it is found that arsenic can remain in the pore water of the clay layer and be released into the aquifer due to overexploitation of groundwater (Smith *et al.*, 2018), ammonium nitrogen released from organic matter decomposition to the methane production stage in a aquitard becomes geological ammonium nitrogen in groundwater (Du *et al.*, 2020), heavy metals in pores of the aquitard have a potential influence to the formation of inferior groundwater (Liu *et al.*, 2020c). Therefore, the hydrochemical characteristics of pore water in low permeability media affect the changes of groundwater components, and the "complex" water–rock interaction under different temperature and pressure conditions is the main cause of diversified inferior components of pore water. The aquitard and topography affects the pattern of groundwater flow systems such as local flow systems, intermediate flow systems, and regional flow systems, and further affect the hydrochemical field including redox, mineralization, temperature, and water chemical type. In groundwater pollution research, aquitard should be considered as "secondary pollution source", in ground subsidence, the deformation of it due to water release also should be considered.

The aquiclude is still a "black box" water-resisting boundary. When studying the chemical causes of groundwater or solute transport, the function of the aquiclude contains great assumptions and uncertainties. At the same time, it is not comprehensive to simply use the aquiclude as the boundary of the groundwater system to distinguish unconfined water from confined water. Based on the Quaternary geological and hydrogeological profile of Jianghan Plain

(a)

(b)

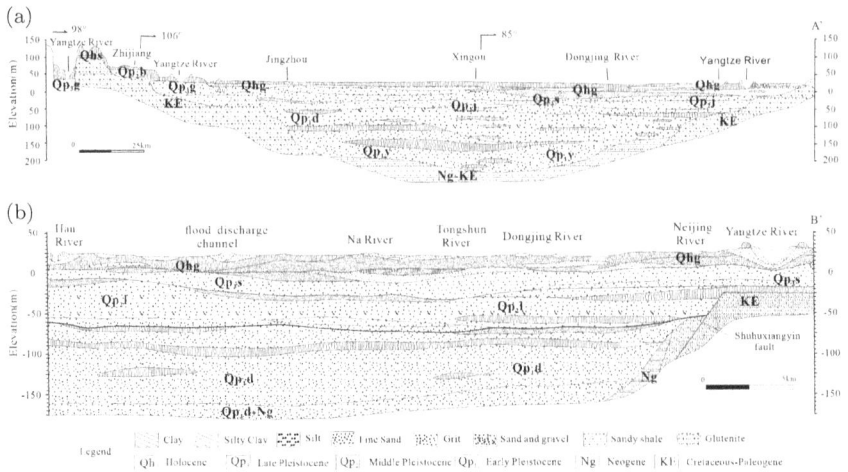

Figure 2.3. Quaternary geological and hydrogeological section of Jianghan Plain (Zhang *et al.*, 2019).

in Southeast China (Figure 2.3), the aquifer (mainly sand layer) and the aquiclude (mainly clay) crisscross in the whole section. There is no regular arrangement and the aquiclude does not cover the whole section, indicating that the confined aquifer in a certain depth is not completely bounded by the aquiclude, and the head of the groundwater is not pressurized where the aquiclude disappears, which makes the distinction between unconfined water and confined water blurred and complicated.

When the groundwater system contains a large-scale ecological environment, due to the complexity of the ecosystem itself, such as being in a large-scale ecological cycle involving water-soil-gas-biology, the boundary of the whole system will expand uncertainly. For example, reclaim land from lakes not only affect the reclaimed area, but also affect the whole lake (Liu *et al.*, 2020d).

2.1.4 *Temporal complexity of system*

The groundwater system, like all-natural systems, has a process of birth, flourish, decadence and death, at different age stages, its performance is different. However, in contemporary research, the groundwater system is generally regarded as a mechanical system without considering its time dimension as a natural system. In this sense, it can be said that the time dimension of the groundwater system is missing.

For example, in the Qilian Mountains, the superposition of multi-stage large alluvial fans occurred in the late Quaternary. These drained alluvial fans, with extremely thick gravel layers, formed steep walls nearly 100 m high, which can be regarded as fossils of the groundwater system. It declares the influence of tectonic dynamics on the birth and death process of the groundwater system. Since the Holocene, the horizontal distance of the left-lateral slip of the fault has reached 55 m, the corresponding rise of the northern plate is 27 m, and the average horizontal slip velocity is $5.5 \pm 2.2 \, \text{mm/a}$ (Peltzer *et al.*, 1988). Such fissure and driven force underground is sufficient to lead groundwater runoff birth, flourish, decadence and death. Tectonic activities in a broad sense are everywhere, with only differences in scale, intensity, and nature.

As a natural system, under the premise of relatively stable recharge conditions, a groundwater system is all in a specific water–rock reaction stage, and in a specific stage of tectonic stress influence. Each system is occurring and developing along with its specific time series. The prospect of exploitable yield, the principle of judgment exploitable yield and overexploitation should be different for systems with different time series.

The complexity of the groundwater system is far more than that. Generally speaking, the groundwater system is random. For example, in the following two profiles of the porous groundwater system in the Yangtze River Delta (Figure 2.3), groundwater with specific runoff patterns and hydrochemical composition will be formed under different spatial locations, sedimentary environments, lithology, or mineral types. Groundwater systems are hierarchical. There are several sandy aquifers and aquitards distributed from shallow to deep, they are stacked on each other and have distinct levels. Also, there are certain levels from the scale of the groundwater flow system (local flow system, intermediate flow system, and regional flow system) (Figure 2.3). Groundwater systems have a time sequence. Some hydrogeological phenomena, or hydrological cycles often have time sequences. For example, the hydrological cycle of small subsystems is faster than that of the whole giant system. Generally speaking, the groundwater system is random. For example, in the following two profiles of the porous groundwater system in the Yangtze River Delta (Figure 2.3), groundwater with specific runoff patterns and hydrochemical composition will be formed under different spatial locations, sedimentary environments, lithology, or mineral types. Groundwater systems are hierarchical. There are several sandy aquifers and aquitards distributed from shallow to deep, they are stacked on each other and have distinct levels. Also, there are certain levels from the scale of the groundwater flow system (local flow system, intermediate flow system, and regional flow system) (Figure 2.3). Groundwater systems have a time sequence. Some hydrogeological phenomena, or hydrological cycles often have time sequences. For example, the hydrological cycle of small subsystems is faster than that of the whole giant system. It also includes a certain time lag between the occurrence of certain substances or phenomena, such as the response of groundwater level to drought or rainstorm (Allen *et al.*, 2010). The groundwater system has self-adaptability. The shallow aquitard often has a certain purification effect, because of its small particles and large specific surface area. It has a strong adsorption capacity for some harmful components, thus reducing the migration risk of pollutants. The groundwater system is fuzziness. The fuzziness of the groundwater system is manifested in many aspects, and in addition to what has been mentioned above, it also includes fuzzy judgment on the real nature of groundwater. We study groundwater by drilling it to the surface and then testing it, some components or properties of groundwater will inevitably change in contact with air, so it is impossible to restore the real properties and components of groundwater as it flows underground.

In a word, when studying complex groundwater systems, we should grasp the common phenomena or laws, find appropriate methods to analyze specific problems. The following will introduce each complex subsystem and its influencing factors in detail.

2.2 Complexity of Medium Space

Rock and soil in nature are all porous medium. There are pores, fissures, or solution cracks of different shapes and sizes in their solid skeletons. Some of them contain water, some do not contain water, and some cannot permeate water although they contain water. Usually, porous media that are both permeable and full of water are called water-bearing media, which is the primary condition for the existence of groundwater. The rock mass that stores groundwater and can flow out of groundwater under natural or manmade conditions is called an aquifer. The material, spatial distribution, and void difference of aquifer will affect the burial, distribution, and movement characteristics of groundwater. The complexity of the water-bearing medium is one of factors leading complex and changeable groundwater system.

2.2.1 *Complex geology*

Among the natural geographical conditions, meteorology, hydrology, geology, and geomorphology have the most significant impact on groundwater. The geological conditions that affect the formation of groundwater are mainly rock properties and geological structures. The nature of rock determines the storage space of groundwater, which is a prerequisite for the formation of groundwater. The geological structure determines whether water can be stored and how much water can be stored. Different geological formations affect groundwater systems differently, so groundwater has different transport characteristics at different geological formations. In tectonic basins, due to the basin-type structure, the basement often deposited the huge thickness Quaternary loose sediments, coupled with good water catchment conditions, good confined aquifers generally formed and rich in artesian water. Affected by fault, concomitant structural fracture in the faulting surface and both sides often strike as the fault, forming the tectonic fracture zone. The characteristics of water conduction-storage of fault are controlled by the lithology of the two plates and the mechanical properties of the fault. When both sides of the fault plane are developed with good tensile torsion fractures, it is a water-conducting fault. The water conduction fault is not only a water storage space but also a water conduction channel and a water collection corridor, connecting several aquifers, so that the water inflow of the water conduction fault developed in the permeable surrounding rock is large and stable. In folded formations, groundwater mainly occurs in a series of secondary structural fissures associated with folding process of the strata. During the formation of the anticline, the axial stress is concentrated, so the anticline generally has good horizontal permeability. Under suitable recharge conditions and lithological association, a large amount of groundwater can be accumulated in the anticline structure to form abundant groundwater resources. The storage and migration of groundwater in folded zone are also related to the nature of the formation itself. In areas with large formation stress would occur interlayer compression and sliding, peeling structures, joint and fracture development, thus large water storage spaces are created (Zhu, 1999).

Except for some crystalline dense rocks, the majority of rocks have certain voids. Pore water exists in pores formed by loose rocks and soil particles and is relatively evenly distributed. Fissure water exists in various fractures formed by internal and external dynamic geological processes in hard rocks, and its distribution is extremely uneven. In some areas where structures are developed and faults are concentrated, rock strata are broken, various fissures are densely distributed, and karst water is concentrated in and around large faults in veins and belts.

The pore water is mainly distributed in pore of loose sediment, widely distributed in the quaternary strata within the piling plains and intermountain basins, and is an important source of water supply for industrial agriculture and living water. Pore water in aquitard is a liquid that is distributed in rock and soil pore or adsorbed by the surface of solid particles, and would not move affected by gravity (Tang and Wang, 1986). The diffusion of solutes is the main mechanism for the transport, transformation, and adsorption of solutes (Hendry and Wassenaar, 2000). In pedology, pore water is defined as a liquid containing a large amount of soluble salts that are maintained on the solid phase surface and in the capillary pores under the suction of soil matrix, it is also called soil solution. Scholars in marine science define pore water as the solution existing in the pores of seabed sediments and rock particles, also known as interstitial water or ooze water. In the disciplines of environmental geology and ecological environment, pore water refers to the liquid adsorbed on the surface of solid particles, also known as absorbed water, pellicular water. In hydrogeology, pore water refers to the liquid that exists in the voids of sediments and is adsorbed by solid particles. Unlike general free liquid water, pore water cannot move freely under its gravity. In general, the porosity in sand and clay is small, the flow velocity of the fluid in them is very slow, the Reynolds number of fluid is small, and the seepage movement law conforms to Darcy's law. However, for coarse-grained porous media (such as gravel and pebble), the experimental results show that when the hydraulic gradient increases to a certain extent, the hydraulic gradient and seepage velocity show nonlinear changes, and the neutral effect of the flow field cannot be ignored. At this time, Darcy's law is no longer applicable (Farmer and Howison, 2006). When the medium is subjected to pressure, deformation is usually accompanied by pore compression and pore seepage, resulting in "pore scale evolution effect" of consolidation deformation. However, for low permeability medium with very small particle size and pore size, particle surface charge will have strong physical and chemical effects on pore water near its surface, thus affecting the seepage characteristics. The influence is closely related to the pore size and gradually becomes significant with the decrease of the pore size. As a result, medium seepage presents a significant "pore scale effect" (Liang, 2010). The structure, pore size, and pressure of porous media will affect flow rate and direction of groundwater seepage, making groundwater seepage variable and complex.

Fissure water mainly occurs in various fissures of rocks. About 20% of the sedimentary rocks on the earth belong to carbonate minerals. Due to diagenesis, weathering, and tectonic processes, fractures of different scales have formed in bedrock. These fractures spread in underground space and communicate with aquifers, forming channels for water conduction and storage. The characteristics of fracture development in bedrock area are as follows: (1) heterogeneity; (2) non-sequential consistency; (3) anisotropy. Part of groundwater exists in limestone pores and fissures and interacts physically, chemically, and biologically with limestone during the flow process, thus expanding the pores and fissures into caves and pipelines.

Groundwater occurring in rock mass fissures can be divided into weathered fissure water, diagenetic fissure water, and structural fissure water according to the genesis of a fissure in the water-bearing medium. The weathered fissure water is usually evenly distributed, and hydraulic links are better, but the water content are limited; the dikes and the contact zone between the intrusive rock and the surrounding rock can form open and ribbon fissures after condensing, where is rich of banded fissure water. During the condensation process of the lava flow, the uncondensed lava flows away, leaving huge lava pores in the rock mass to form tubular water-bearing zones, which can become water-rich layer. Structural fissure water is characterized by uneven distribution and poor hydraulic connection. According to the burial conditions, it can be divided into unconfined or confined water. Compared with pore water, fissure water is unevenly distributed, the hydraulic connection is not good, and the permeability of the medium is heterogeneous and anisotropic. In most cases, the movement of fissure water conforms to Darcy's law. Only in a few huge fissures does the movement of water not conform to Darcy's law and even belongs to turbulent movement. The important difference between the fractured medium and porous medium is that fractured medium has heterogeneity and anisotropy. Lithology, size, opening degree, density, direction, and distribution of fractures all affect the occurrence and movement of fissure water, resulting in the complexity of the groundwater flow system. Lithology affects the enrichment degree of fissure water by different filling materials and filling degrees in fracture zones. Metamorphic rocks and clastic rocks are mostly plastic strata, with poor fracture opening and often tightly filled with argillaceous materials. The water content and water conductivity of fractures are greatly reduced. However, igneous rocks such as granite and granodiorite are brittle rocks with good fracture opening, less filling materials, and most materials are sandy, which make them have relatively good water content and water conductivity. Reticular fissure channels are often developed on the surface of the formation formed by weathering. This kind of fissure is small and can easily accept precipitation supply. Vein-like and banded fissure channels are generally developed at the axis and turning of fold structures with concentrated in-situ stress, they are controlled by geological structures. Longitudinal fissures developed in the same direction as the main structural line and transverse fissures perpendicular to the main structural line are formed under the action of tension. Generally, they have a good opening, deep cutting, and long extension and are the main water conduction channels. Tensile and tensile–torsional fractures are the most water-rich fractures.

The occurrence space of karst groundwater is a karst water-bearing medium. As a complex aquifer system, the karst water-bearing medium contains a variety of forms such as pores, cracks, karst pipes, and small caves etc. The uneven spatial distribution of water richness and anisotropy of hydraulic links are essential characteristics of karst water-bearing medium. With the circulation of groundwater in a limestone area, the complex and changeable water–rock interaction produces a large area of geological phenomena such as dissolution, deposition, precipitation, subsidence and underflow, ponor, and karst spring. In hydrogeology, the land-form produced by interaction of water and soluble rocks in different stages is called karst. The dissolution of fissures and rock layers in limestone areas is the driving force of early karstification (Matter *et al.*, 2007). When water containing CO_2 flows into a limestone fissure, H_2O–CO_2–$CaCO_3$ forms a triple system where the chemical equilibrium and reaction dynamics laws within the system determine the evolution of the limestone fissure.

Given the inhomogeneity of the karst water-bearing medium, Swiss scholar Bogli (1980) believes that the basic characteristics of the karst water-bearing medium are pipeline flow and inconsistent groundwater level. American scholar White (1988, 2002) thinks that the water-bearing medium of fractured carbonate rocks consists of intergrain pores, fissure pores, and pipelines and quantitatively calculates the proportion of the three types of pores.

Different geological tectonic movements will form different water-bearing media, and different media will have different effects on groundwater storage, recharge and discharge.

2.2.2 *Complexity of climate and landscape*

Global climate change will affect the spatial and temporal distribution of atmospheric precipitation in local areas. Atmospheric precipitation is an important part in the hydrological cycle. Its changes directly affect the runoff of surface water and groundwater recharge in the basin. Extreme precipitation events will lead to hydrological disasters such as floods and droughts in the basin (Piao *et al.*, 2010). The impact of climate change on groundwater remains largely unknown because climate systems are intricate, characterized by complex interactions and feedback networks. Generally speaking, in many places, the amount and intensity of precipitation are expected

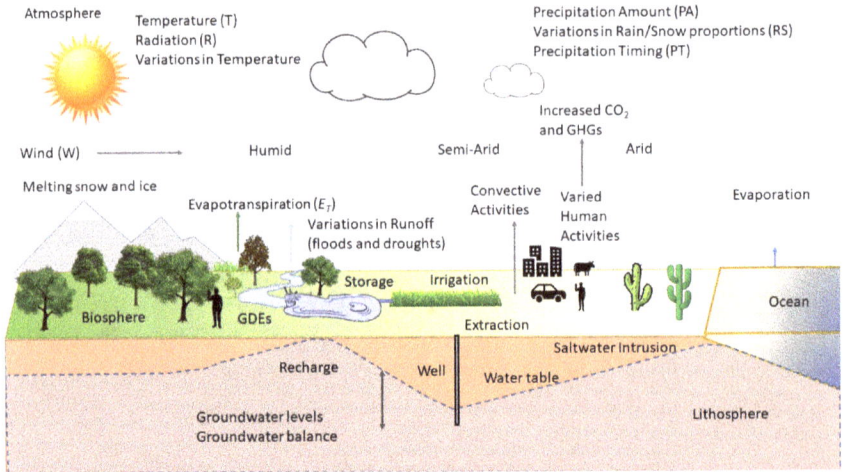

Figure 2.4.   Interaction between the groundwater system and earth components under climate change (Amanambu *et al.*, 2020).

to increase, while others are expected to experience drought. Also, snowfall is likely to decrease. The decrease in snowfall accelerates melting, leading to an increase in average annual surface runoff, especially in temperate climate areas. Figure 2.4 describes the complex interactions between groundwater and the components of the climate system.

Until recently, there was few research on the relationship between climate change and groundwater. The uncertainty involved in detecting the nature and characteristics of subsurface water explains to some extent why understanding the response of groundwater to climate change develops so slowly. In addition, groundwater is relatively immune to seasonal or even decade-long climate change. Given that the effects of climate change are often influenced by human factors and indirect factors such as land-use change and groundwater overexploitation, a comprehensive understanding of groundwater responses to climate change is even more challenging.

Atmospheric precipitation is the main source of recharge of groundwater, and the amount of precipitation directly affects the abundance of groundwater in an area. In humid areas, the rainfall is heavy, so that the surface water is abundant, and the amount of groundwater is also abundant; in arid areas, the rainfall is light, so that the surface water is poor, recharge of groundwater is limited, and the amount of groundwater is generally small. Besides, there is strong evaporation in arid areas, shallow groundwater is often concentrated, combined with poor recharge and circulation, the high-mineralization groundwater would form.

Global warming makes frequency of global high temperatures and droughts is increasing. Rising temperatures lead to an increasing evaporation. In some areas, especially in agricultural irrigation areas, surface water resources alone cannot meet the water supply requirements in dry years, which prompt human beings to overexploit groundwater resources, making the groundwater extraction faster than the recharge, thus the water level declines, leading groundwater depression cone and ground depression etc. Many studies have shown that climate drought plays a positive feedback to the decline of groundwater level. Another direct consequence of global warming is that the water cycle will undergo major changes. With changes in precipitation and evaporation, global groundwater resources will be redistributed. Climate change not only affects the recharge and circulation of groundwater but also affects the water–rock interaction through the changes of rainfall chemical composition and surface temperature, further affects the groundwater quality.

In addition to the effect of precipitation, climate change may also affect groundwater recharge through changing groundwater use. Excessive groundwater extraction, which is mainly used for irrigation, may have a significant impact on groundwater, as irrigation water accounts for nearly 60–70% of all freshwater extraction. Generally speaking, using groundwater irrigation leads to the depletion of groundwater storage, but in some areas, extensive irrigation recharge have been observed. Ou *et al.* (2018) predict that irrigation recharge will steadily increase till 2100 due to increased pumping. In areas where surface water irrigation dominates, irrigation recharge is also expected to increase, but the net effect of over-irrigation is groundwater depletion. With the increase of population and food demand, and the development of the economy, irrigation-led groundwater extraction may even become the most important way for climate change to affect groundwater. In addition, land use/land cover can alter groundwater recharge. Many studies have shown that the replacement of natural vegetation by cropland or building leads to significant changes in groundwater recharge. For example, cutting down forests for agriculture reduces the leaf area, that can increase groundwater recharge, even if rainfall decreases slightly. Groundwater recharge decreases due to increased vegetation density, such as changes from grassland to forest land (Oliveira *et al.*, 2017) or due to replacement of building of rapid urbanization. In general, land use/cover changes, whether the changes are temporary (vegetation change) or permanent (urbanization), can affect the recharge of groundwater by changing the moisture balance (evaporation, transpiration, and surface runoff processes). These additional impacts make it difficult to assess the impact of climate change on groundwater.

There are two feedback mechanisms of groundwater to climate, one is the contribution of groundwater to sea level rise. Groundwater would discharge into the ocean directly or become a part of surface water firstly by extraction, then flow into the ocean. Another is increasing evaporation. Surface energy balance may be broken if adding soil water content by Irrigation with groundwater, this may produce feedback on precipitation. To better understand the feedback of groundwater to climate, it is necessary to describe in detail the groundwater–surface water interaction in the land surface hydrological stage of the atmospheric circulation system.

Under different topographic and geomorphological conditions, formed groundwater is different. Geomorphology is often linked with geology and then with the hydrogeological background of the groundwater system. In plain and basin areas with flat terrain, loose sediments are thick, and the surface slope is small, the surface runoff formed by precipitation is slow in velocity, which can easily infiltrate into the ground and recharge groundwater, especially in coastal areas and the south where precipitation is heavy, under this condition, the groundwater formed is generally low in salinity. Surface water and groundwater in plain areas are closely related. In addition to the recharge by precipitation, surface water can also recharge groundwater, where is mainly concentrated along rivers and around lakes. However, groundwater also has a strong recharge effect on surface runoff during periods of low flow.

In desert areas, despite the ground material is coarse and the water is easy to permeate, the arid climate and light precipitation

still make little recharge of groundwater, resulting in highly mineralized groundwater. In mountain areas, the terrain is steep and the
bedrock is exposed. Groundwater mainly exists in fissures of various rocks and shows a unevenly distribution. And the precipitation has a vertical distribution with altitude, resulting in abundant
groundwater recharge. Different landforms have different effects on
the preservation, flow, and mineralization of groundwater, creating
the complexity of the groundwater system.

2.2.3   *Interaction of earth system structures*

Groundwater belongs to the earth's hydrosphere, and the lower
boundary of the hydrosphere is also the boundary of deep groundwater. The complex interaction between the hydrosphere and atmosphere, biosphere, terrestrial sphere, and the inner sphere of the earth
also constitutes the complexity of the groundwater system.

The hydrosphere is the product of the earth's geological history development and is formed and developed with the changes
of space and time of the earth's crust. Therefore, spatially, surface
water and groundwater are different forms of water but interact with
each other, the unity of them should be considered, including the
unity of groundwater and the relevant water-bearing medium, and
the unity of hydrosphere formation and geological history should be
considered temporally. The terrestrial hydrosphere can be divided
into two parts: shallow hydrosphere and deep hydrosphere in the
vertical direction. The shallow hydrosphere includes surface water,
unconfined water, and shallow confined water. The deep hydrosphere
includes deep confined water, deep highly mineralized water, high-
temperature hot gas and water etc.

The main characteristic of the shallow hydrosphere is that the
above-mentioned three kinds of water have a very close connection with each other, with frequent mutual transformation, interaction, and significant mutual influence. Modern climate are the
common controlling factor for them. The main characteristics of the
deep hydrosphere are that the formation, distribution, and movement of groundwater are mainly controlled by geological structural
conditions, and modern climatic have no significant effect on them.
Because deep groundwater has a long history of formation, a wide

range of recharge and discharge, and is located in the deep part, the alternating speed of water is slow, the hydrochemical reaction is complex, and the circulation period is long, so the groundwater often has large head pressure and high mineralization. In the deep hydrosphere, the interaction between water and the water-bearing medium is of great significance, and they are closely related in the formation history.

The dynamic changes of groundwater in the hydrosphere are not only affected by atmospheric precipitation, recharge, and discharge etc., but also affected by regular changes in the crust and lithosphere. Under shallow buried, groundwater and soil water are closely linked and frequently transformed and the groundwater is affected by plant growth and human activities.

The driving forces of change in the earth system come from earth's interior, solar radiation, and human activity. Surface water and groundwater in the hydrosphere enter the atmosphere through evaporation, and enter the biosphere through plant and human activities. Water vapor in the air returns to the biosphere through rainfall and snowfall. Groundwater circulates continuously in the earth system and is jointly affected by property of lithosphere, atmospheric and biosphere. These complex influencing factors constitute groundwater's complexity.

2.3 Complexity of Hydrodynamic Field

2.3.1 *Influence of medium structure complexity on hydrodynamic field*

The flow of groundwater mainly depends on two factors, flow channel and driving force. When both are available, groundwater will flow. Since groundwater moves in the voids of soil and rock media, the speed of movement is affected by the permeability of aquifer media. The stronger the permeability of rock strata, the less blocked the groundwater movement is, and the easier the groundwater movement is. The difference in groundwater level is an important driving force for its movement. The greater the difference of water level per unit distance, the stronger the driving force, and the easier it is for groundwater movement. Groundwater is divided into pore water, fissure water, and karst water according to medium types.

The flow of pore water in Quaternary loose porous media conforms to the law of conservation of mass and Darcy's law. Its seepage has the following characteristics: (1) the surface area per unit volume of pores in porous media is relatively large and the surface effect is obvious. Stickiness must be considered at all times; (2) the pressure is often high in underground seepage, so the compressibility of fluid is usually considered; (3) the pore shape is complex, the flow resistance is large, capillary force is common, and sometimes molecular force should be considered; (4) it is often accompanied by complicated physical and chemical processes.

Since Darcy's law was established, after 150 years of hard work, groundwater seepage mechanics have made great progress. However, there are still many important issues that need to be continuously studied. Such as (1) the scale effect of parameters is the basic theoretical content closely related to porous medium theory and continuous medium mechanics. It is directly related to the mechanism of groundwater seepage, mass transfer, and heat transfer and determine the practical application effect of seepage theory; (2) seepage in unsaturated medium due to the nonlinear characteristics of seepage in unsaturated in rock and soil mass, not only does the permeability coefficient depend on the volume water content of the soil but also there is a moving front, so the research on this has always been one of the difficult contents in groundwater seepage mechanics; (3) seepage in the non-continuous medium. The seepage of groundwater in the pore requires further study.

Due to the heterogeneity of fracture spatial distribution, groundwater seepage and solute transport in fracture medium also have strong heterogeneity, even are multi-dimensional and multi-scale. Regarding the water seepage characteristics in rock mass fissures, many scholars have used the water seepage laws of broken rock mass or post-peak fractured rock mass to describe the water flow characteristics in water-conducting fissures. In addition to the compression-shear cracks after the rock strata are damaged by plastic yield, there are also a large number of tension cracks caused by the rock strata breaking and rotating movement in the water-conducting fissures of the rock mass. There are obvious differences in the spatial distribution characteristics and development forms of the two in the rock mass, and there are essential differences of the water flow state in the tensile-shear zone and the caving zone. Therefore, it is difficult to fully and accurately reveal the conductivity characteristics of different fractures of rock mass simply by taking broken or post-peak fractured rock mass as the research object.

In the study of fissure water seepage, the existing test results show that the permeability coefficient of the rock mass is inversely proportional to the motion viscosity coefficient of fluid, and the motion viscosity coefficient of fluid is a function of temperature. The temperature field affects the distribution of the seepage field by changing the permeability coefficient of the rock mass, and the temperature gradient will also affect the motion of water flow. Many works have done to study the related problems. Such as the study on influence of thermoelastic stress on fracture seepage showed that under the action of thermal expansion of rock mass, fractures self-close and promote fluid movement (Xi *et al.*, 1998); the discrete element method has been used to study the coupling problems of water and heat exchange in fractured rock (Abdallah *et al.*, 1995).

The movement characteristics of karst water are affected by the void network of the karst system. The greater the heterogeneity of the void (fissure-cave) network, the greater the difference in water movement in the fissure-cave network. The movement characteristics of groundwater in the karst system include laminar flow and turbulent flow coexisting, confined and unconfined coexisting, inconsistent movement direction, inconsistent water level. For underground pipelines or large connected pores, fluid flow equations can be used for good simulation.

The characteristics of karst water movement depend on the shape and size of the voids in carbonate rocks. The voids in carbonate rocks can be divided into two types: substrate voids and cleavage voids with caves. Although limestone and dolomite look hard, they have fine tectonic cracks and native pores under the microscope, the porosity of them could reach to 9.5%. The water in the substrate voids does not only communicate with each other through micro-fissures but also communicates with the water in the fissures. Therefore, substrate voids are important storage spaces for karst water. When the carbonate rock mass is located below the regional drainage datum level, the fissure water will continuously move, dissolve the rock and widen the fissure, and will finally form karst caves. The karst cave controls the movement of karst water, making a karst water system with a unified water level. Karst fissure can be generally divided into karst cave fissures, micro-fissures, and substrate voids according to their sizes. From the viewpoint of hydraulic action, karst media can be divided into three categories: (1) water storage media, which is composed of substrate voids and plays a role in water storage; (2) the water conveyance medium consists of cracks and fissures, and its function is to provide water passage; (3) the water control medium partially or completely spreading in the karst system. The horizontal karst cave controls the movement of karst water and makes it a curved surface of groundwater level. Therefore, the karst medium is called the triple medium.

The zoning of the karst groundwater is mainly composed of an upper-layer interstitial water circulation zone, interstitial unconfined water circulation zone, and deep interstitial water infiltration zone. The karst water circulation models can be divided into the following types: monocline, reverse monocline, strike, syncline basin, fault block etc., in which the interstitial unconfined water circulation zone is the main water storage zone, and the deep interstitial water infiltration zone is the recharge system of deep groundwater or confined water. Different evolution models represent different water yield characteristics, and their development are usually along the direction of structural zones or only occur in weathered zones.

The flow of pore water is mainly affected by lithology, medium structure, and pore size. The flow of fissure water is mainly controlled by the development of the fracture and its spatial distribution. The flow of karst water depends on the shape and size of the carbonate voids. Therefore, the dynamic fields of these three types of water are all affected by the complex structure of the water-bearing medium.

2.3.2 *Influence of heterogeneity on hydrodynamic field*

Different particle compositions, lithologic minerals, and diagenesis determine the heterogeneity of aquifer, which controls groundwater flow and solute migration characteristics. The hydraulic conductivity is the most important parameter in characterizing the heterogeneity properties of aquifers (Lavigne *et al.*, 2010) and one of the most important parameters in the simulation of groundwater flow and solute transport. Among all the uncertain factors affecting underground flow, the heterogeneity of the aquifer permeability coefficient is an important factor for the uncertainty of groundwater movement.

Aquifer heterogeneity has a significant impact on the groundwater flow system and is an important factor controlling the groundwater flow model. At present, the research of the influence of aquifer heterogeneity on water flow system is mainly limited to the theoretical research under the conceptual model, and most of them are assuming that the permeability coefficient is distributed regularly (Freeze *et al.*, 1967). Under actual conditions, the spatial variability of aquifers is complex, especially in aquifers formed in an alluvial–diluvial environment, due to the existence of a large number of randomly distributed small viscous and silty viscous lenses, the heterogeneity of aquifers has the characteristics of randomness and layered structure. Because it is difficult to determine the position and size of lenses one by one, aquifer heterogeneity is difficult to accurately describe, which often leads to the distortion of basin groundwater simulation and limits the understanding of the formation and evolution law of groundwater flow system.

Taking riverbed sediments as an example, relevant studies show that under the condition of relatively small hydraulic gradient change on the bed surface, the characteristic variables of hyporheic exchange are mainly controlled by the heterogeneity of riverbed sediments. Sawyer and Cardenas (2009) analyzed the influence of riverbed sediments with cross-bedding on hyporheic exchange, revealing that sediment heterogeneity not only affects the flow movement track entering the hyporheic zone but also affects the residence time in the hyporheic zone. Pryshlak *et al.* (2015) further revealed that sand wave and curved reach jointly affect the flux magnitude and residence time distribution model on the water–sediment interface. In terms of flume tests, Salehin *et al.* (2004) constructed different sediment structures in the flume, and the effects of strong and weak heterogeneity sediments on hyporheic exchange are compared, revealing that relatively strong heterogeneity can lead to large hyporheic exchange flux at the water–sediment interface. This basic conclusion has been further verified in the field experiment study in India Creek, Southeast Pennsylvania, USA. Blois *et al.* (2014) studied the influence of weak permeable riverbed and high permeable riverbed on hyporheic exchange through flume tests, respectively. The study revealed that high permeable riverbed sediments can affect the pressure gradient distribution on the bed surface, and upwelling in the hyporheic zone will affect the surface turbulence characteristics. It can be seen that the heterogeneous permeability of groundwater medium can affect the movement process of groundwater, but the heterogeneity of medium is difficult to describe, which will lead to the complexity of viscous force distribution and driving force strength of groundwater in a heterogeneous medium.

2.3.3 *Complexity of external factors*

Groundwater is the main component of the water cycle, and its dynamic changes are closely related to human activities and climate change. The main driving factors that affect the groundwater level and runoff in the region are groundwater recharge and exploitation, and climate change and human activities are the root causes of the main driving factors change. With the increasing exploitation of groundwater, the balance of groundwater recharge and discharge is destroyed, resulting in the groundwater level declined year by year. At the same time, the chances of precipitation, evapotranspiration, and unconfined water evaporation caused by climate change also affect the relationship between groundwater recharge and discharge. Since the middle of the 20th century, driven by the global warming, the frequency of extreme weather such as heavy rainfall and abnormal high temperature has increased, coupled with the unreasonable exploitation and utilization of water resources by human beings, has had many impacts on the hydrodynamic field. For example, a series of hydrological environmental responses such as groundwater level fluctuation, groundwater runoff, and discharge have attracted worldwide attention. Climate and human activities will change the distribution of hydrodynamic field, but this change is not single but is jointly changed by multi-factor coupling. The change of hydrodynamic field will also change with the climate or human activities. The complexity of external conditions affects the complexity of the hydrodynamic field. The root causes for the changes in the main driving factors are described as follows:

1. *Climate change*: Climate change affects the hydrodynamic field by affecting the groundwater level, water content, recharge, and discharge.

Water flowing from underground to surface, evaporating to the atmosphere, or extracted by human, all represent groundwater discharge. Five major processes of groundwater discharge have been identified: (i) spring water flow; (ii) transpiration of local vegetation; (iii) evaporation of soil and open waters; (iv) underground outflow; and (v) extraction for various human uses.

Climate change is mainly the change of climate factors. The main climate factors that affect the dynamic change of groundwater are temperature, precipitation, and evaporation (Figure 2.5). Under the background of global climate change characterized by climate warming, soil evaporation and vegetation transpiration enhance with the increase of temperature, so that groundwater evapotranspiration consumption increases, resulting in groundwater discharge increase. Precipitation is the basic element of the watershed water cycle and the main source of groundwater resources. During the flood season, precipitation increases, surface runoff increases, and groundwater recharge increases. At the same time, groundwater exploitation decreases and groundwater level rises. On the contrary, during the drought period, precipitation decreases, surface runoff decreases, and groundwater recharge decreases, resulting in an increase in groundwater mineralization in shallow groundwater. At the same time, groundwater exploitation increases and groundwater level decreases. Excessive evaporation may cause rainfall to convert into vapour before it can recharge groundwater, which is not conducive to groundwater recharge. Evaporation has an upward trend, would weaken groundwater recharge and the increase drought degree, leading to the decline of groundwater level, causing ecological and environmental problems such as soil salinization and land desertification and large-scale overexploitation so that changing groundwater reserves and water level dynamics.

Atmospheric precipitation penetration is an important source of groundwater recharge, and there are many studies on the impact of precipitation on groundwater. Brunke and Gonser (1997) summarized the impact of precipitation on groundwater and river water, arguing that the base stream is the discharge of groundwater to river water when precipitation is light; when the precipitation is heavy, the runoff of surface water and groundwater increases, and the hydrodynamic field would change, that the river water would recharge the groundwater, especially in flood seasons. And this recharge might be

(a)

(b)

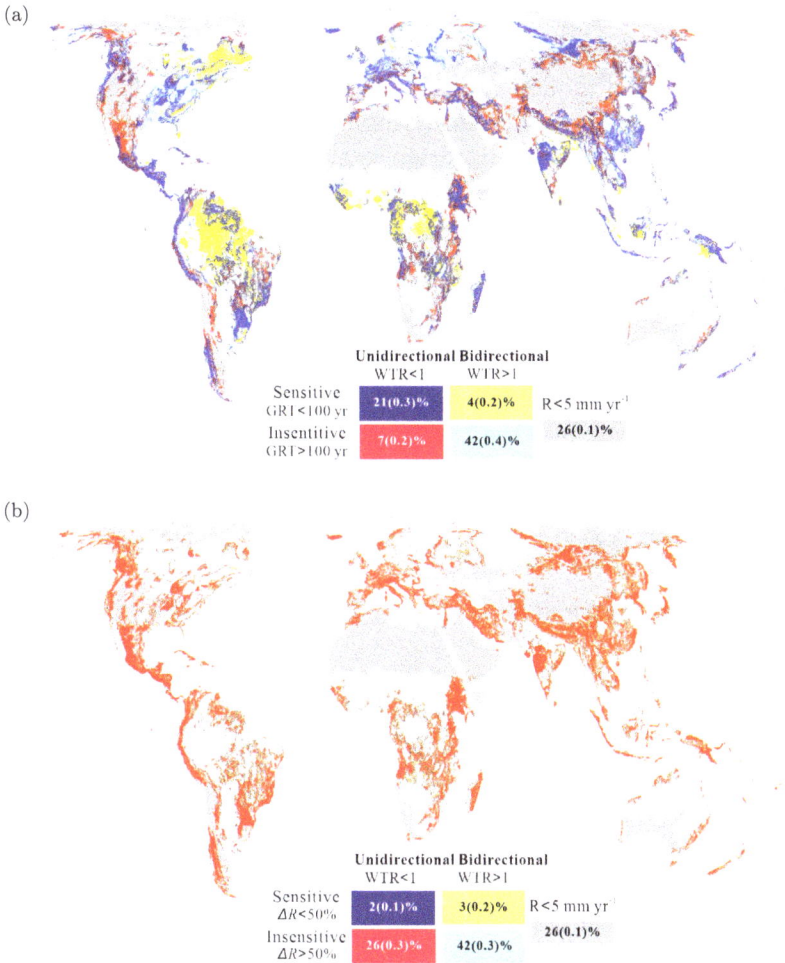

Figure 2.5.   Global distributions of the temporal and spatial sensitivity of the mode of climate-groundwater interactions. (a) Temporal sensitivity; (b) Spatial sensitivity. (WTR is the water table ratio, which is a measure of the relative fullness of the subsurface and thus the extent of the water table's interactions with topography. GRT is the groundwater response times. R is the hyper-arid regions of recharge.) (Cuthbert *et al.*, 2019).

controlled by duration of flood, terrain, permeability of riverbed and water storage of river bank.

The dynamic change of groundwater level can cause change in the hydrodynamic field and change the relationship between groundwater recharge and discharge. To study the influence of climate change on groundwater dynamics, scholars have applied various methods and models (Scibek and Allen, 2006). A large amount of observational evidence shows that the world is currently experiencing obvious climate change due to the increase of greenhouse gas concentration in the atmosphere (Kirshen, 2002). As previous studies have shown that climate change and groundwater extraction may be the main factors contributing to changes in groundwater levels, it is important to study the evolution of groundwater in long-term climate change. Extreme precipitation is a special case of precipitation, and its effect on groundwater is relatively complex. The increase of extreme precipitation also makes it easy for a large number of substances to infiltrate into the ground with runoff, resulting in a greater risk of groundwater pollution. Therefore, how to make rational use of extreme precipitation, put it to the ground, and prevent groundwater from pollution is also a key content of future research.

2. *Human activities*: The hydrodynamic field is complex. The temporal and spatial distribution and evolution of hydrodynamic field are not only restricted by natural factors but also affected by the social environment, especially human activities. Human activities have various impacts on land water reserves, such as reservoir or dam regulation, agricultural irrigation, as well as industrial and domestic water consumption. The significant drop in groundwater level caused by high-intensity human exploitation, has affected the water cycle processes such as precipitation infiltration law, water use in crop root layer, vertical recharge of aquifer, and groundwater level dynamics to some extend. Only understanding the basic characteristics of the water cycle process with large buried depth can the water resources in relevant areas be reasonably utilized (Owuor *et al.*, 2016).

The influence of human activities on the hydrodynamic field of groundwater systems is mainly reflected in two aspects, that is, recharge and discharge. In the aspect of groundwater recharge, due to the interference of human activities such as mining, and drainage, the groundwater level dropped significantly. After that, the precipitation infiltration process was prolonged, the precipitation infiltration recharge coefficient became small, and the water infiltration recharge amount decreased significantly under the same precipitation situation. Changes in land use patterns also have an impact on groundwater recharge. In some areas, the irrigation area has expanded rapidly, the irrigation water consumption has made the irrigation leakage recharge increased accordingly, further changing the groundwater recharge structure. In terms of groundwater discharge, the rapid increase of groundwater exploitation has changed the structure of groundwater discharge, and the natural discharge has been captured is the main cause of ecological environment problems such as the reduction of base flow, the depletion of spring water, and the shrinkage of wetlands. Taylor and Stefan (2009) have found that the groundwater temperature in developed urban areas with high land utilization rate located in the Minneapolis–Sao Paulo regions of the United States is higher than that in areas with low land utilization rate at the same latitude. To expand the space, the world's coastal areas solve the severe "land deficit" problem through reclamation, mainly including salt pan development, aquaculture, port-centered industry, offshore oil field exploitation, and coastal tourism. Affected by these human activities, the balance of land water and seawater exchange has been broken. Groundwater recharge, runoff, and discharge conditions have been altered, causing seawater intrusion and groundwater salinization etc. Reclamation activities have changed the hydraulic connection among surface water, soil, and groundwater systems and changed the groundwater environment, the mechanism of groundwater quantity and salt transport has become a difficult point in current research.

The impact of water conservancy projects on groundwater is also complex, on the one hand, water system connection projects can effectively compensate groundwater; on the other hand, connection projects will also change groundwater circulation. The influence of river dam construction on the downstream surface water environment and ecology has been widely studied, but the research on the changes of groundwater environment closely related to surface water under the influence of dam construction is still very lacking. Most of the studies are only qualitative descriptions and inferences of the dam's impact on downstream groundwater.

Human activities will cause changes in the hydrodynamic field and change the hydraulic connection among surface water, soil, and groundwater systems. Groundwater recharge, runoff, and discharge conditions would be changed, resulting in changes in the groundwater system. However, how the hydrodynamic field changes, the amount of groundwater migration, and the mechanism of water–rock interaction under the influence of human activities still need to be studied urgently.

2.4   Complexity of Hydrochemical Fields

The chemical composition of groundwater is affected by recharge sources, runoff characteristics, surrounding rock properties, and their mutual transformation with surface water and precipitation. The groundwater buried very shallow may show different degrees of salinization, while the groundwater buried deep reflects the strength and distance of runoff. The hydrochemical fields exhibit horizontal and vertical zoning evolution and can be divided from upstream to downstream into freshwater zones, freshwater–microbrackish zones, and brackish water zones. In aquifers of complex sedimentary systems, the solid medium may be various rocks that have been cemented, such as sandstone, siltstone, limestone, dolomite, and gypsum layer or uncemented loose sediments, such as gravel stone, sand, and clay. These sediments contain a variety of minerals, and the mineral complex so the solute source of groundwater is complex. The formation and influencing factors of groundwater chemistry are complex, the most important factors are the following: first, the spatial sequence of water–rock interactions; second, the type of biogeochemical processes in groundwater systems.

The groundwater is mainly derived from atmospheric precipitation, followed by surface water. Before entering the aquifer, these waters obtain certain substances from the atmosphere or other media. After entering the aquifer, they continuously interact with rock and soil, further changing their chemical composition. There are the following seven processes, leaching, concentration, decarbonization, desulfurization, cation alternating adsorption, mixing, and human activities, which promote the evolution of the chemical composition of groundwater. Leaching is a common effect, which makes rock and soil lose some substances and groundwater adds some new components. Water is composed of a negatively charged oxygen ion and two positively charged hydrogen ions. Due to the asymmetric distribution of hydrogen and oxygen, one end near the oxygen atom is negative to form polar molecules. When water interacts with rock and soil, charged polar molecules of water often capture ions with weak binding force in mineral lattices into the water, and leaching causes components in rock and soil to enter groundwater. Leaching is a widespread water–rock interaction in nature. The chemical composition of groundwater is closely related to the products of water–rock interactions and reflects groundwater and mineral interaction processes, including mineral corrosion, mineral precipitation, cation exchange, and the secondary SiO_2 solubility and CO_2 content in aqueous solutions etc. (Wang *et al.*, 2016).

Groundwater will contact with different rock masses or soils during the flow process, which will cause different chemical reaction processes. For example, layers of sandstone, greystone, shale, and gypsum exist in groundwater flow systems, and the order in which water meets them is different. The assumed first encounter sequence is limestone (containing calcite), gypsum layer ($CaSO_4 \cdot 2H_2O$ or $CaSO_4$), sandstone (containing quartz and feldspar), and shale (containing sodium montmorillonite); the second sequence is sandstone, shale, limestone, and gypsum layer. Then, the final result is different, the first one may form SO_4-Na type water, and the second might form SO_4-Ca type brackish water.

Although the rocks encountered in the groundwater flow system are the same, groundwater with different chemical characteristics might form due to different orders of encounter. Therefore, the formation and evolution of the chemical composition of groundwater in complex sedimentary systems should pay particular attention to the encounter with rocks in the groundwater flow system in practical information analysis.

The change of hydrochemical field is mainly affected by natural and human factors. Natural factors are groundwater recharge and water–rock interaction. The influence of human factors on the chemical field is mainly manifested in two aspects: on the one hand, excessive extraction of groundwater changes the runoff conditions of groundwater and the migration and transformation process of some solutes; on the other hand, industry, agriculture, and daily life discharge a large amount of wastewater, garbage, and other wastes, causing pollutants to enter the ground and change the hydrochemical field (Chen *et al.*, 2008), so complex impact factors create a complex hydrochemical field.

2.4.1   *Coupling factors of media and dynamic field*

The groundwater media field is a complex system in the presence of groundwater and flood sediments, lake sediments, shore-delta sediments, karst, glacial sediments, etc. The driven force for groundwater movement is different in varied media fields, so the groundwater movements are varied, which may have several influencing factors such as geological conditions, temperature, potential energy, water properties, and media field.

The dynamic field of groundwater refers to pressure during groundwater movement or fluid potential energy including gravity, inertia, and viscous force. The mechanism of groundwater movement includes compaction fluid effects, hydrothermal effects, osmotic effects, and clay dehydration etc. For example, the hydrodynamic field in Songliao Basin is composed of multiple hydrodynamic sub-systems, and its formation and evolution have obvious asymmetry on the plane. Generally speaking, the northern part of the basin is a centripetal flow area of atmospheric precipitation infiltration. The central depression area is the centrifugal flow area and discharge by cross-formational flow area. The southern part of the basin is mainly characterized by groundwater cross-formational flow and concentration by evaporation, and the edge of the basin and some areas in the uplift denudation area are atmospheric precipitation infiltration areas, which form local hydrodynamic units with characteristics of formation pressure, flow direction, fluid potential, vertical pressure gradient, etc. At the same time, the formation and evolution of hydrodynamic field have stages, and its hydrodynamic intensity has zoned vertically. From shallow to deep, it can be divided into three zones: strong, weak, and stagnant. The hydrodynamic system takes the depression of the basin as the center and the edge of the basin or the adjacent uplift ridgeline as the boundary. From the edge of the hydrodynamic system to the center of the depression, centripetal flow of precipitations and/or cross-formational flow and centrifugal flow develop in turn (Figure 2.6) (Lou *et al.*, 2006).

Geological structure and hydrogeological conditions play a decisive role in the formation of groundwater chemical composition to a certain extent. These are mainly manifested in the differences of topography, strata properties, groundwater circulation conditions, and burial, and the coupling of medium field and dynamic field.

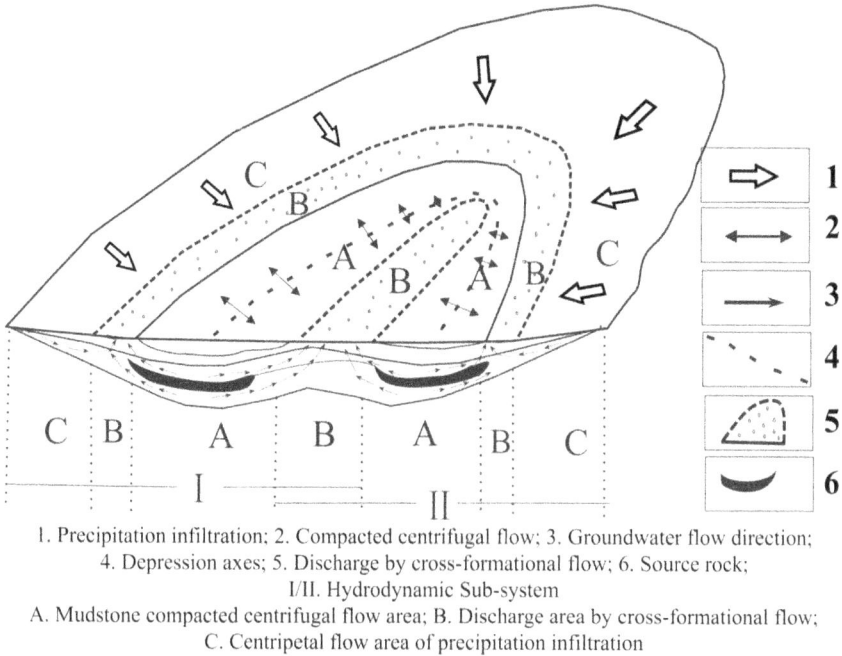

1. Precipitation infiltration; 2. Compacted centrifugal flow; 3. Groundwater flow direction;
4. Depression axes; 5. Discharge by cross-formational flow; 6. Source rock;
I/II. Hydrodynamic Sub-system
A. Mudstone compacted centrifugal flow area; B. Discharge area by cross-formational flow;
C. Centripetal flow area of precipitation infiltration

Figure 2.6. Diagram of relationship between local hydrodynamic unit and hydro-dynamic system in sedimentary basin (Lou *et al.*, 2006).

When studying groundwater pollution (including DNAPLs/ LNAPLs), the coupling effect of groundwater medium field and hydrodynamic field has a great indicating effect on solving hydro-chemical environmental problems. For example, when the concentration of pollutants in adjacent high permeability zone is relatively high, these low permeability zone can be used as tanks for pollutant storage first. After pollutants are removed from the high permeability zone, the pollutants would enter the high permeability zone again from low permeability zone by reverse diffusion. At this time, low permeability zone would be secondary pollution source. Reverse diffusion means that after pollutants accumulate in large quantities in LPZs (low permeability zones), solute transfers from LPZs to HPZs (high permeability zones), and the concentration gradient reverses. This mode was first proposed by Wilson (1997) and was observed by Ball *et al.* (1997). At present, field research mainly focuses on the

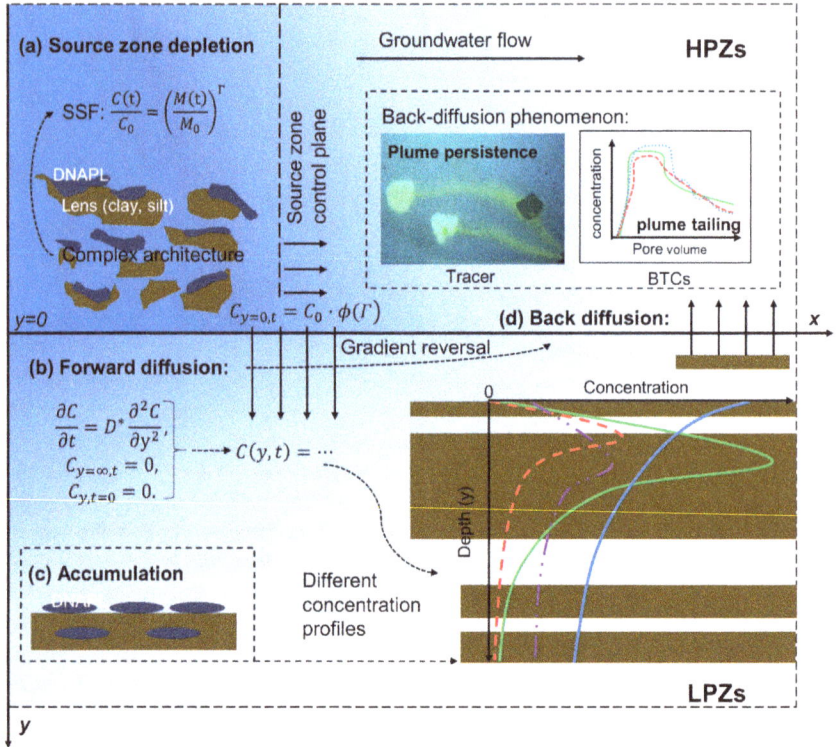

Figure 2.7. Concept model of the contaminant occurrence and migration between HPZs and LPZs (You *et al.*, 2020).

characterization of hydrogeological properties and the collection of pollutant concentration data in high and low permeability zones, to demonstrate the persistence and reverse diffusion of pollution plumes in specific locations (Weill *et al.*, 2017). The three main steps of this are as follows: (1) the dissolution of the pollutant during forwarding diffusion; (2) the forward diffusion and accumulation of contaminants in LPZs; (3) the reverse diffusion effect of LPZs (You *et al.*, 2020) (Figure 2.7).

2.4.2 *Biogeochemical processes*

The redox conditions of groundwater systems are highly variable, with many influencing factors mainly including the type and quantity of oxidizing agents and reducing agents in groundwater, the cycle process of groundwater, as well as the types of microorganisms and organic matter contained in groundwater (Christensen *et al.*, 2000). Due to the differences in temperature, acidity or alkalinity between river water and groundwater, there are obvious physical gradient, chemical gradient, and biological gradient in the infiltration and recharge area of river water, thus a strong and complicated biogeochemical process takes place (Yuan, 2017). Microorganism can reproduce both in unconfined water and deep groundwater (even more than 1,000 m). Most of the important chemical reactions in groundwater, especially those containing organic matters and redox processes, are completed through the microorganism. There are various microorganism in groundwater, including nitrifying bacteria, sulfur bacteria etc., in oxidizing environment, as well as desulfurizing bacteria, methanogenic bacteria, ammoniation bacteria etc., in reducing environment (Zhang *et al.*, 2005). Microorganism activities in the groundwater environment have a great influence on the chemical composition of groundwater, rock surface characteristics, and material migration. At the same time, there are many kinds of terminal electron acceptors in the formation (mainly O_2, NO_3^-, Fe (III), Mn^{4+}, and SO_4^{2-}). They directly affect the distribution of microbial population and microbial activities (Thomas *et al.*, 2001). Oxidation capacity is an index to evaluate the oxidation ability of the underground environment, and its value mainly depends on the number and effectiveness of terminal electron acceptors in the formation. The environment with low oxidation capacity is easy to form zones with low redox levels (such as methane-producing zones), while the environment with high oxidation capacity will limit the formation of methane-producing zones. Therefore, oxidation capacity is an important parameter to evaluate biogeochemical processes and pollutant attenuation in the underground environment (Christensen *et al.*, 1994). In a groundwater environment, microorganisms consume (reduce) dissolved oxygen and NO_3^-, Fe (III), SO_4^{2-}, and CO_2, oxidizing organic matter to obtain energy. This process is closely related to elemental geochemical cycle in the groundwater system. For example, in dissolved oxygen and NO_3-reduction stage, the groundwater is in a oxidation environment, at which the reduction of Fe (III) is inhibited and the arsenic loaded by Fe (III) will not be released into the groundwater. When Fe (III) is reduced, the groundwater is in a reducing environment, which will lead to the release of arsenic coexisting with Fe (III) and form high arsenic groundwater. When SO_4^{2-} is in reduction, the groundwater is in a strong reduction environment, and the iron sulfide formed by HS^- and Fe (II) will adsorb or coprecipitate arsenic, which will reduce the arsenic concentration in the groundwater (Zhang *et al.*, 2015). Also, in the methane-producing zone, iron reduction zone, NO_3^- reduction zone and the oxygen reduction zone, the oxidation capacity increases in turn, while the reduction capacity decreases in turn, and these adjacent zones are partial overlapped (Dong *et al.*, 2006). The above changes all affect the biogeochemical reaction and the migration and transformation mechanism of pollutants in groundwater.

Many studies have confirmed that groundwater and vegetation are mutually dependent and influenced. Groundwater level and salinization degree are the main factors affecting vegetation distribution and growth, and they also determine the difficulty of saline-alkali land improvement directly. The greater the salinization, the greater the negative impact on vegetation growth (Xu, 2016). Meantime, vegetation can also conserve water sources and maintain groundwater level, and change the hydrochemical characteristics of groundwater. For example, high transpiration in vegetation would lead to an increase in unconfined water salinity, selectively absorbing ions by vegetation would change the groundwater hydrochemical characteristics. Besides, the decomposition of vegetation residues and root effect etc., also could change the groundwater quality by leaching of soil.

2.4.3 *Human activities*

With the rapid development of the economy and society, human activities are exerting important influence on nature. For example, human life and production activities (such as the extraction of groundwater, the construction of water conservancy, the development of irrigation, mine drainage, artificial recharge, and excessive use of chemical fertilizers and pesticides) will lead to the increase of groundwater hardness and pollution and change groundwater chemical types, thus having a great impact on the formation and evolution of groundwater. On the one hand, manmade wastes would pollute groundwater; on the other hand, the formation conditions of groundwater are artificially changed on a large scale, thus changing the hydrochemical properties of groundwater.

The excessive use of groundwater resources by humans will cause a drop in groundwater levels, which further changes the groundwater recharge, runoff, and discharge conditions near the mining area. Excessive exploitation will also reduce the water output of production wells and even lead to deterioration of water quality. Artificial exploitation will have different impacts on unconfined and confined groundwater. For unconfined groundwater, (1) the redox conditions of vadose zone would be changed due to unconfined groundwater level decline, the components from vadose zone would change the hydrochemical characteristics of groundwater; (2) the groundwater would be salinized due to the recharge of adjacent seawater, salt lake water or saline groundwater. For confined groundwater, (1) if regional cone of depression occurs, the lower saline groundwater would get into the target aquifer by cross-formational flow so that changes the hydrochemical characteristics; (2) the pressure within the depression decreases, CO_2 in the water would escape when groundwater flows from high pressure to low pressure, indirectly causing the carbonate precipitation of iron and alkali metals, thus changing the chemical composition of groundwater; (3) the hydrostatic pressure of the aquitard and the overlying strata decreases due to the decrease of the confined water level, resulting in an increase in the effective pressure of the clay layer and a compaction effect, so the water in the clay would enter the aquifer and change the chemical composition of the groundwater.

Human activities cause the release of pollutants, these pollutants might get into the aquifers, changing the hydrochemical field and causing groundwater pollution. Groundwater pollution is harmful to human health and industrial and agricultural production. There are some obvious differences between groundwater pollution and surface water pollution, because pollutants enter the aquifer and move slowly in the aquifer, pollution often occurs gradually and is difficult to detect in time without special monitoring; after groundwater pollution is found, it is not as easy to determine the pollution source as surface water. More importantly, groundwater pollution is not easily eliminated. After removing pollution sources, surface water can be purified in a short period. However, for groundwater, even if pollution sources are removed, pollutants that have entered the aquifer will still have long-term adverse effects. These are some typical groundwater pollution cases, the excessive exploitation of underground freshwater leads to the intrusion of sea (salt) water in coastal areas; the nitrate pollution caused by surface sewage (waste) water discharge and agricultural pollution; pollution from petroleum and petrochemical products; landfill leakage pollution. Among them, agricultural pollution is characterized by a large amount pollutants and a wide range distribution.

Human engineering activities can also affect the natural running of groundwater hydrochemical field such as the construction of water conservancy project. The construction of water conservancy projects not only regulates surface runoff and local climatic conditions, but also changes the runoff condition of groundwater by affecting the infiltrated amount of surface water. In addition, during the mining, the groundwater levels would continue decline with mining scale increase, resulting in change of recharge, runoff and discharge condition of groundwater. These changes might influence the recharge source and recharge area, even transfer the discharge area to recharge area, which often occurs in karst areas. The processes would change the hydrochemical field.

Water injection project would carry out to reduce disaster like ground depression caused by overexploitation of groundwater. It injects surface water into the aquifer with or without pressure, physicochemical property of groundwater changes while regional groundwater level rises again (Bouwer, 2002). A corresponding test of the water quality variation in the artificial recharge has been conducted. The test showed that main components in water varied non-significant. Although Ca^{2+} and HCO_3^- varied with time, the variation is limited, sometimes even less than the analysis errors. For example, Ca^{2+} and Mg^{2+} content change in the groundwater was generally only 1–4 mg/L, and there was only one increase of 8 mg/L during the stop-recharging period. However, other components such as Fe^{2+}, Mn^{2+}, DOC, and pH changed significantly over time (Yan *et al.*, 1987).

To sum up, the above activities can be summarized into the following three types: human overutilization of groundwater resources; the release of pollutants caused by farmland, factories, and other living and production activities; large-scale water conservancy projects have changed the runoff process of the groundwater system. These anthropogenic activities can alter the chemical characteristics of groundwater, such as increased hardness of groundwater, increased levels of pollutants, and altering the type of groundwater chemistry and thus have an important impact on groundwater formation and evolution.

2.5   Stress Field and Thermal Field of Groundwater System

With the development of social economy, the exploitation of underground space and deep underground resources has increased, such as the exploitation of deep groundwater, oil, natural gas, some deep mineral resources, high dams for hydropower projects, water drainage tunnels, road and railway tunnels and deep underground treatment of highly radioactive nuclear waste. These underground projects are located in the rock mass simultaneously with heat, seepage, and stress fields, when the underground environment is disturbed, these fields would interact each other, eventually reaching a state of dynamic equilibrium. Due to the complexity, the coupling of thermal field, seepage field, and stress field of rock mass in groundwater system has become a hot and difficult topic (Pan *et al.*, 2009; Rutqvist *et al.*, 2008; Oda *et al.*, 2002).

2.5.1   *The complexity of the stress field*

The change of rock mass stress state will cause the change in medium seepage parameters, thus affecting the flow of groundwater and changing the fluid migration channel and water chemical field. This is the influence of the stress field on the seepage field. On the contrary, groundwater in the rock mass system can change the structure of the rock mass through physical, chemical, and mechanical actions and make the pressure of the rock mass change. At present, the coupling effect between the stress field and seepage field is mainly studied from three aspects: porous medium, fractured medium, and anisotropic rock mass.

It is found that under the coupling action of thermal field, seepage field, and stress field, the initiation, expansion, and connection of rock mass cracks caused by engineering disturbance are the main factors causing the failure of such engineering rock mass. For example, deep underground nuclear waste repositories, due to the heat release of nuclear waste decay and the stress redistribution of deep surrounding rock caused by excavation, it is easy to cause the initiation and expansion of cracks in the surrounding rock of the nuclear waste repository, thus changing the stress field and seepage field of the surrounding rock, thus safety accidents such as the destruction of the surrounding rock of the nuclear waste repository and the leakage of nuclear waste would happen (Tsang *et al.*, 2009). As another example, for mining engineering, with the increase of mining depth, the rock mass will be under harsh environments such as high ground stress, high temperature and high water pressure. In the case of excavation disturbance, the surrounding rock stress will redistribute, the surrounding rock temperature will change due to the air introduced, and the pore water pressure of the surrounding rock will also change due to variation of the recharge, runoff, and discharge conditions of groundwater caused by excavation. The changes of stress field, thermal field, and seepage field of surrounding rock occur at the same time, which easily causes crack initiation and propagation of surrounding rock, then the instability failure of roadway surrounding rock would occur. Therefore, when analyzing the coupling problem of thermal field, seepage field, and stress field in the fractured rock mass, it is necessary to focus on the expansion and connection of fractures in rock mass under the condition of multi-field coupling.

However, the traditional numerical simulation method cannot simulate the expansion and evolution of the fracture network in the multi-field coupling of the fractured rock mass properly. For example, when the finite element method and the finite difference method are used to deal with the crack propagation problem, the mesh needs to be re-divided in each step of crack propagation, which greatly increases the calculation workload and singular phenomena will appear at the crack tip of the stress field. The extended finite element method (EFEM) simulates cracks by introducing an additional function reflecting the discontinuity of cracks into the shape function (Belytschko and Black, 1999). However, this method has some difficulties in simulating complex crack propagation.

The study of fluid–solid coupling seepage began in the 1940s. The theoretical models of seepage in a fractured rock mass can be divided into three types: seepage model of equivalent continuous medium, seepage model of fracture network discontinuous medium, and seepage model of two-phase medium. The equivalent continuous medium seepage model is based on the seepage tensor theory, which simplifies fractured rock mass into continuous medium uniformly and describes seepage problems with the continuous medium method. When the actual velocity of fluid in the rock mass fissures is much larger than the Darcy velocity in the equivalent continuous medium, the fissure network discontinuous medium seepage model or the two-phase medium seepage model should be adopted at this time. The seepage model of fracture network discontinuous medium regards the fractured medium as a network system that fissures of different sizes and directions crossing in space. Water can only move in the fracture network, while the movement can be regarded as one-dimensional flow in a series of parallel pipelines. The seepage model of a two-phase medium assumes that the rock mass is a continuous medium with overlapping porous medium and fractured medium. Porous medium stores water and fractured medium conduct water. By studying the water exchange process between the two medium and the deformation of the two medium, the coupling calculation can be carried out by introducing Biot's porous elastic media theory and single fracture cubic law. Jabakhanji and Mohtar (2015) derived the seepage model of porous media based on the near-field dynamics theory.

However, natural fractures have complex material composition, structure, and physical and mechanical characteristics, especially the hydraulic characteristics under high stress are very complex, which undoubtedly brings great difficulties to numerical simulation. At present, the research on the coupling of seepage and stress in a single fracture is far from satisfactory. In the aspect of numerical simulation, special attention should be paid to the change of mechanical properties of fractures under a complex stress environment. The rock mass will undergo plastic deformation, be crushed, or milled under a complex stress environment, and micro-fractures may also occur in rock blocks near the fracture surface. Therefore, the coupling of seepage and stress in a single fracture should not be simply modified the cubic law, but should further consider the influence of plastic deformation of the fracture surface, groove flow phenomenon, and some random factors. The change of seepage field in the fractured rock mass and the effect of groundwater (including chemical corrosion, physical weakening effect, and mechanical effect) will cause corresponding changes in a stress field environment. It leads to seepage deformation of the fractured rock mass. This deformation process has certain timeliness, which is manifested in the following: (1) the physical and chemical effects of groundwater on fractured structural planes gradually weaken the physical and mechanical properties of fractured rock mass; (2) groundwater expands the structural plane in fractured rock mass through mechanical action.

2.5.2 *Complexity of thermal field*

The change of groundwater temperature relies on conduction and convection, while convection is based on the seepage effect of groundwater, so when seepage occurs, heat will be transferred in the form of convection to redistribute and groundwater temperature changes. Due to the existence of groundwater and the vertical temperature gradient of the stratum, the change of the aquifer seepage field can be sensitively reflected in the groundwater temperature. Furthermore, the change of the stress field that causes the change of the seepage field can be indirectly reflected in the change of groundwater temperature. At present, the research on the relationship between groundwater seepage field and thermal field is mainly carried out from the perspective of the hydrodynamic mechanism, such as the relationship between groundwater vertical movement, flow rate, velocity and water temperature (Feng *et al.*, 2015). These studies can reveal the relationship between the change of rock mass seepage field and the change of groundwater micro-thermal field caused by stress during earthquake preparation and occurrence.

In most cases, the temperature of groundwater depends on the mixing ratio of deep hot water and shallow cold water. When the amount of hot water mixed increases, the groundwater temperature increases. On the contrary, when the inflow of hot water decreases, the groundwater temperature decreases. The anomaly of groundwater temperature in well or hot springs is generally believed to be related to mechanical processes. On the one hand, when fissures and faults as hot water supply channels are opened or closed due to stress, the inflow of hot water will inevitably increase or decrease, and the groundwater temperature will rise or fall. On the other hand, stress

Table 2.1.   Density of pure water at different temperatures.

Temperature/°C	Dens/g/mL	Temperature/°C	Dens/g/mL
0	0.9999	40	0.9922
4	1.0000	50	0.9881
10	0.9997	60	0.9832
20	0.9982	80	0.9718
30	0.9957	100	0.9584

can also lead to changes in water pressure, thus causing changes in flow rate, which can also cause changes in hot water inflow and water temperature. For cold wells, the abnormal fluctuations of water temperature is usually the result of mixing of groundwater in different aquifers caused by stress and may sometimes be related to the upwelling of deep hot water (Zhou *et al.*, 2013).

The influence of temperature on groundwater seepage field is mainly reflected in the following two aspects:

1. *Influence of temperature on groundwater density*: The volume of water with the same mass is different at different temperatures, which is manifested by the change of density. As general, groundwater density reaches the top at 4°C, which is $1\,\mathrm{g/cm^3}$. With the increase temperature, the density will gradually decrease, as shown in Table 2.1. However, the influence of temperature on groundwater density is less than other factors, so it can be ignored under certain situation.

2. *Effect of temperature on the viscosity of groundwater*: The effect of temperature on the viscosity of groundwater can be estimated from the Helmholtz empirical formula (Montazeri *et al.*, 2014), which is as follows:

$$\frac{\eta}{\eta_0} = \frac{1}{1 + 0.03368T + 0.0002209T^2} \tag{2.1}$$

η_0 in the formula is the viscosity of groundwater at 0°C, and η is the viscosity of groundwater at temperature T.

Figure 2.8 shows the relationship of groundwater viscosity and temperature when $\eta_0 = 1.7921$ cP. The viscosity of groundwater will greatly decrease with the increase of temperature, indicating that

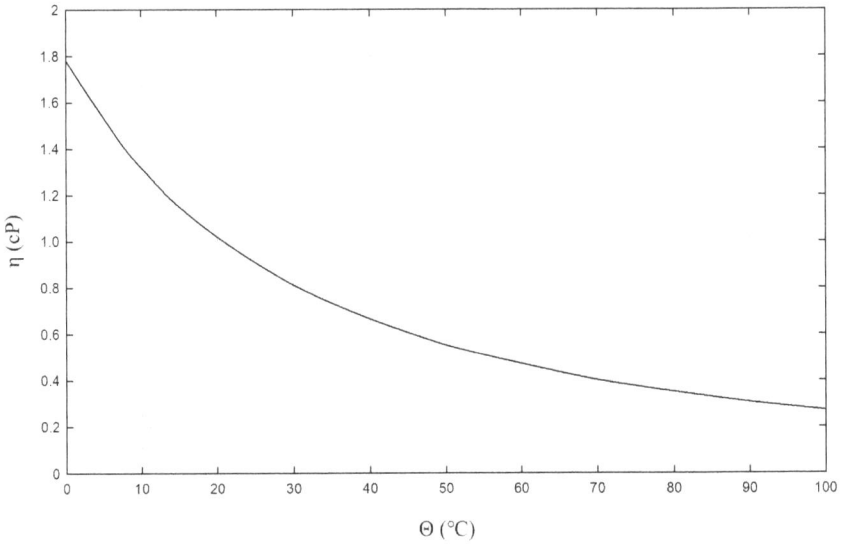

Figure 2.8.   Relationship curve between groundwater viscosity and temperature (Montazeri *et al.*, 2014).

the change of groundwater temperature will have a great impact on its viscosity, which would affect the permeability coefficient of groundwater in the rock mass, and further affect the seepage field.

In addition to the physical complexity, the chemical changes that are affected by thermal and stress fields in the water-bearing medium are complex, either. The low permeability medium mentioned above evolved mainly from mud in areas with large ancient lake landforms such as the Jianghan Plain. In the process of burial evolution, mud is continuously covered by new sediments, which will lead to compaction and diagenesis. In the initial stage, pores are continuously compressed, pore water is continuously discharged, and mud will gradually evolve into a clay layer. After that, with the further burial process, the clay will eventually evolve into mudstone (or shale). Generally speaking, the burial process of mud can be divided into two types, physical process and chemical process. The physical processes are mainly mediated by pressure. However, the chemical process is much more complicated, mainly mediated by microorganisms in shallow burial periods and mainly mediated by temperature in deep burial periods (Potter *et al.*, 2005).

The physical process of mud–clay evolution is mainly the compaction and consolidation of mud under a load of overlying water bodies and sediment. The most significant effect of this process is the reduction of mud porosity and the massive discharge of pore water. When the buried depth is less than 500 m, pore water is easily discharged and porosity decreases sharply. When the buried depth is more than 500 m, the decrease of porosity slows down significantly, because after a large amount of pore water is discharged, the continuous decrease of porosity depends on the discharge of interlayer water and structural water closely combined with clay materials. A large number of field geological surveys show that when the buried depth of the strata is more than 500 m, argillaceous sediments generally exist in the form of clay rocks. Therefore, the process of mud–clay evolution mainly releases pore water and bound water.

The chemical process of mud–clay evolution is a complex water–rock interaction process mediated by microorganisms, which will roughly undergo the evolution of oxygen-containing zone, oxygen-free zone, and methane-producing zone from top to bottom. Each with its corresponding microbial population is aerobic bacteria, sulfate-reducing bacteria, and methanogens (Ortega-Guerrero *et al.*, 1993). In the top layer of mud, free oxygen is continuously consumed by the organic matter under the action of aerobic bacteria. When oxygen is exhausted and becomes an anaerobic environment, sulfate reduction becomes the main chemical process. When sulfate is almost completely reduced, methanogens can convert CO_2 to CH_4 (Judd and Hovland, 2007).

The evolution process of the mud to clay aquitard can be regarded as a complex water–solid–gas interaction process in which temperature and pressure gradually increase and transition from the oxidation environment to the reduction environment. At the same time, mud will release a large amount of water in the process of evolving into a clay aquitard. In the process of recharging the adjacent aquifer with the water released from the mud, chemicals originally enriched in the mud will also enter the aquifer with the release of the pore water. Since clay pore water is enriched in heavy metals and other contaminants relative to aquifer groundwater (Wang *et al.*, 2016), a large amount of pore water entering the aquifer will largely affect the evolution of water quality in adjacent aquifers. Also, it is worth noting that chemical components in the aquitard in the solid or adsorbed form are also continuously interacting with adjacent aquifers, further affecting the evolution of groundwater quality. Thus, it is of great significance to carry out in-depth research on this process for enriching relevant theories of hydrogeology. However, due to the extremely complex evolution process of mud–clay, few studies have linked mud and clay aquitard.

2.5.3   *Coupling of stress field and thermal field*

Aiming at the coupling problem of temperature-seepage-stress in groundwater system, years of research have found that the fractures in the rock mass are usually filled with water. The fracture opening degree of the rock mass would change due to the external stress and fluid pressure, then the hydraulic conductivity seepage field of the rock mass would change. This will also cause a change in the stress field further. At the same time, the flow of water changes the distribution of the thermal field in the rock mass. The thermal field, seepage field, and stress field are continuously coupled during this deformation of the rock mass (Long *et al.*, 2011).

To reveal the mechanism of groundwater temperature change under stress loading, in addition to the macroscopic study of thermal field changes caused by groundwater migration, it is also necessary to study the mechanism of evolution of porous medium and fissure media widely existing under stress as well as the migration law of fissure fluid, and further study the heat migration and thermal field change law of fissure medium. The fissure medium acts as a special vehicle for stress loading, fluid transport, and heat field changes, a great deal of research work has been done on its force evolution, hydrodynamics, and heat transport laws (Ma *et al.*, 2013). The exploration of groundwater heat transport in porous medium began in the 1960s, and some progress has been made in experimental research, theoretical research, and calculation methods. However, due to the high complexity of heat migration in fractured media, the geometric factors of fracture space, permeability parameters, dispersion coefficient have significant uncertainties, which makes it extremely difficult to study heat migration in the fractured rock mass. At present, the research level of water and heat migration in pore medium greatly exceeds that in fractured medium.

The processes causing a temperature change in the tiny real unit are as follows: (1) convection of heat with water particles; (2) heat conduction by liquid and solid media; (3) mechanical dispersal during heat transport. Besides, there are factors such as natural heat convection and heat exchange between water and rock caused by uneven temperature. At present, the research that only considering the influence of one of the factors accounts for the majority, and few reports comprehensively consider the above factors to simulate and calculate the heat transport.

The change of groundwater thermal field is an objective response to tectonic stress loading to some extent, but it is not easy to truly recognize the quantitative relationship between them in theory. In the study of groundwater variation during earthquake, many studies focused on seepage field changes under the stress, such as observe well water levels and flow changes, and the simulation results also can well correspond to the observations. Water temperature changes reflect the stress changes just as sensitively as water level changes, even more responsive than other observations. However, research work on groundwater temperature changes is few, due to the complexity of changes in groundwater thermal field. Using the existing theory of the relationship between the stress field and the seepage field to further study the thermal field change, will be the main method to reveal the groundwater temperature change under stress loading.

Chapter 3

Uncertainty of Groundwater System Observation and Forecast

3.1 Parameter Uncertainty

3.1.1 Spatial–temporal variability of media field and hydrodynamic field

Hydrogeological parameters are important quantitative indicators of the hydrogeological performance of water-bearing medium. In general, hydrogeologists use hydrogeological testing methods (e.g., pumping test, water injection test, infiltration test, and Darcy's test), groundwater dynamics observations (wiring method and local uniformity method) to identify the groundwater system flow patterns, and hydrogeological parameters (Ding *et al.*, 2015). The determination of water-bearing system parameters is of decisive significance to the accuracy of regional groundwater system observation and prediction, while the medium field and hydrodynamic field of groundwater system have spatio-temporal variability, which greatly increases the difficulty of water-bearing system observation and prediction. Tectonic movements (lithospheric deformation, ocean floor accretion, earthquakes, etc.) inside the earth, gravitational effects (solid tide, ocean tide, etc.) generated by planets around the earth, and some meteorological factors will all affect the hydrogeological parameters (Elkhoury *et al.*, 2006; Yan *et al.*, 2016). Also, the hydrogeological parameters of the aquifer will change as the completion time of the monitoring well.

Therefore, it is of great significance to determine the temporal and spatial variation and influencing factors of medium field and hydrodynamic field for clarifying the uncertainty of groundwater observation parameters and avoiding the influence of parameter uncertainty on the observation and prediction accuracy of the groundwater system.

3.1.1.1 *Spatial–temporal variability of the media field*

The characterization of the groundwater system media field depends on the determination of the parameters of the water-bearing system. Water-bearing system parameters are indexes that reflect the hydrogeological performance of the aquifer or permeable layer, mainly including porosity, permeability coefficient, water release coefficient, and water storage coefficient, which are described as follows:

1. *Porosity*: The porosity of rock and soil determines the water storage capacity of rock and soil and also controls the ability of rock and soil to retain, release, and transport water under certain conditions (Figure 3.1). Porosity (n) is usually used to represent the proportion of pores in a unit volume of rock and soil. In nature, the primary factor affecting geotechnical porosity is sorting, followed by particle arrangement, particle shape, and pore size (Zhang *et al.*, 2011). The porosity of clay soil is much larger than that of coarse-grained soil because the surface of clay particles is charged. During the deposition, clay particles aggregate to form overhead aggregates, which can form pores larger than the particle diameter.

There are mainly three methods to calculate the effective porosity of rock and soil mass. One is based on the relationship between

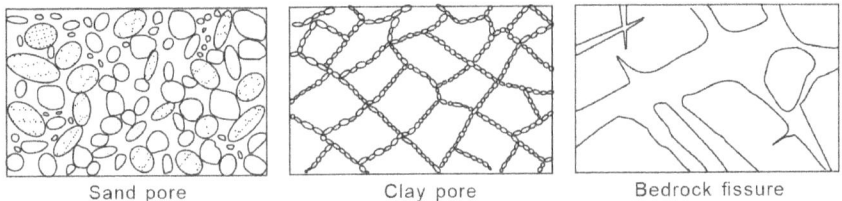

| Sand pore | Clay pore | Bedrock fissure |

Figure 3.1.   Porosity of rock and soil (Revised by Zhang *et al.*, 2011).

equivalent porosity and water suction. The second is based on the pore size distribution and total porosity measured by the mercury intrusion method on the premise of setting the minimum permeable pore radius. The third method is to use tracer experiments to obtain the actual average velocity and dispersion coefficient of groundwater in the field and then to obtain it according to the convection–diffusion equation (Du *et al.*, 2002).

There is an obvious positive correlation power function relationship between porosity, permeability coefficient, and water storage coefficient. In a unit volume aquifer, permeability coefficient and water storage coefficient increase with the increase of porosity. The hydraulic conductivity coefficient is the product of the permeability coefficient and the aquifer thickness, so there is also a power function relationship between porosity and hydraulic conductivity coefficient (Ding *et al.*, 2015). It can be seen that the temporal and spatial changes of rock and soil porosity will affect the other hydrogeological parameters in the water-bearing medium.

Under natural conditions, the porosity of rock and soil will change affected by the tectonic movement. The drastic fluctuation of groundwater level caused by human activities will affect the stress (effective stress) borne by the geotechnical skeleton, thus affecting the porosity. Mining confined water or semi-confined water will lead to the decrease of piezometric level, then the pore water pressure will decrease, resulting decrease of porosity from compression of geotechnical skeleton.

2. Permeability coefficient: Permeability coefficient (K), also known as hydraulic conductivity coefficient, is an important parameter for evaluating permeability in the hydrogeological field. In an isotropous medium, which is defined as the unit flow at a unit hydraulic gradient, denotes how difficult it is for fluid to pass through the pore. The permeability coefficient is expressed in the tensor form in anisotropic medium. The greater the permeability coefficient, the more permeable the aquifer. The coefficient of permeability of the aquifer itself is primarily controlled by the nature of rock formations and certain physical properties of the fluid (such as glutinousness). The permeability coefficient of formations with different textures is quite different. For example, the empirical K value determined by the field pumping test of the silty layer is 0.5–1 m/d, while that for gravel layer is 80–125 m/d (Table 3.1).

Both natural and human factors would affect the permeability coefficient. Natural factors mainly include tectonic movement and tidal action inside the earth, that change the permeability coefficient by changing stratigraphic lithology. For example, the seismic wave increases the permeability coefficient of the aquifer by dredging the fissures, opening the fissures, and forming new fissures. Human factors mainly affect the accuracy of the pumping test through well flushing etc., thus affecting the obtained permeability coefficient.

At present, the commonly used permeability coefficient measurement methods include on-site pumping tests and indoor permeability tests. Scholars at home and abroad have also summed up a large number of empirical formulas (Wang *et al.*, 2013) for calculating permeability coefficient through parameters such as characteristic particle size, pore ratio, and porosity. The limitation of boundary conditions in indoor testing, the difference between the depth of the pumping test in the field and the impact scope of the project, would result in difference between the calculated permeability coefficient and the actual value. The permeability coefficient obtained from the field pumping test is usually less than that calculated in the laboratory (ideal conditions), and the difference between the two is large (Table 3.1). The main reason is that actual aquifers are more non-homogeneous and obtained permeability coefficients may be mixed of multiple aquifers, which cannot have ideal structural conditions

Table 3.1.   Difference in empirical values of permeability coefficient (Sin *et al.*, 2019).

Lithology	Empirical value of pumping test (m/d)	Laboratory empirical value (m/d)
Gravel pebble	80–125	>200
Sandy gravel	45–100	100–200
Coarse sand	20–50	25–50
Medium sand	20	10–25
Medium fine sand	5–17	8–18
Fine sand	1–8	5–10
Silt	0.5–1	1–5

consistent with the laboratory. The results of the pumping test, although derived from a large amount of actual information, are affected by the system and random errors, so the accuracy of the results will also vary from the true value of the aquifer.

During the pumping experiment, drilling methods and water flushing will affect the permeability coefficient to some extent. If mud drilling is used, and the impact of mud blocking around the well is not cleaned thoroughly by water flushing before the pumping experiment, the permeability coefficient calculated would be smaller than those using water drilling. The reason is that during the drilling process, not only the mud itself forms a mud wall but also the mud will seep into the pore in the aquifer around the well, greatly reducing the permeability of this layer (Xin *et al.*, 2019). Also, during the pumping test, the hydraulic jump inside and outside the well pipe and the deviation of the flow moves from the straight line when the flow pattern changes from laminar flow to mixed flow, will affect the measurement of permeability coefficient.

3. *Water release coefficient and water storage coefficient*: Water release coefficient and water storage coefficient are important hydrological parameters. When the water head of the aquifer drops, the deadweight pressure acting on the top plate or phreatic surface of the aquifer increases relatively, forcing the aquifer to compress vertically and part of pore water to be extruded. This phenomenon is called the water release of the aquifer. The water release coefficient is the amount of water (water layer thickness) released by vertical compression deformation of the aquifer column per unit cross-sectional area when the water head or water level drops per unit depth, which is equal to the compression deformation amount of the column, including elastic water release coefficient and plastic water release coefficient. Correspondingly, the water storage coefficient represents the amount of water stored by the aquifer column per unit cross-sectional area due to rebound expansion. Since only elastic compression deformation can rebound, although the concepts of elastic water release coefficient and water storage coefficient are different, they are the same in quantity (Qiu, 1989).

In hydrogeological exploration, hydrogeological data are often obtained by a single well unsteady flow pumping test, and water release coefficient and water storage coefficient are determined by

Theis formula (Singh, 2006). Due to instruments or human factors, there will be certain errors in the observation of water level drop depth. When the water level drop value with errors is substituted into the Theis formula to calculate hydrogeological parameters, the formula itself will transfer the errors of the original data. The expression for the Theis formula calculates the hydrogeological parameters as:

$$H = \frac{Q}{4\pi T} W\left(u\right) \tag{3.1}$$

$$W(u) = \int_\alpha^\infty \frac{e^{-\alpha}}{\alpha} d\alpha \tag{3.2}$$

$$u = \frac{r^\alpha \mu'_s}{4Tt} \tag{3.3}$$

$$A = -e^u W\left(u\right) \tag{3.4}$$

where, H — Depth of water level drop

Q — Pumping volume

T — Hydraulic conductivity

r — Distance from the calculation point to the pumping well

t — Time variable

μ'_s — The elastic water release coefficient of the aquifer (approximately in quantity equal to the water storage coefficient).

When $|A|$ is greater than 1, the Theis formula itself magnifies the error of water level degradation and narrows when $|A|$ is less than 1. From formula 3–4, it could be seen that the absolute value of A is related to the size of u, when $u > 0.4348$, $|A|$ is less than 1 and decreases with u values increase, and when $u < 0.4348$, $|A|$ is greater than 1 and increases with u values decrease (Guo, 1989).

In practical work, the aquifer hydrogeological parameters are used to calculated by simplified Theis formula when $u < 0.01$, and in this case, the relative error of hydrogeological parameters is at least four times of that for water level drop, so when applying the Theis simplified formula to calculate the elastic water release coefficient, a high accuracy of water level drop is needed. In the pumping test, the water level drop depth is sometimes observed by the pumping well, because u decreases with the decrease of r. When the pumping time is slightly long, the u value on the borehole wall will become very small, and the water level drop depth observed by the pumping well is often large, which brings large errors to the calculation of the elastic water release coefficient.

Therefore, to improve the accuracy of the calculation of the elastic water release coefficient, the water level drop at the later stage of pumping should be avoided as much as possible, and observation wells should be kept as far away from pumping wells as possible.

3.1.1.2 *Spatial–temporal variability of hydrodynamic field*

The hydrodynamic field of the groundwater system is primarily characterized by aquifer hydrogeological parameters and motion elements (hydraulic slope, flow rate, etc.) (Chen *et al.*, 2011). The formation and evolution of hydrodynamic field is the combination the comprehensive reflection of combination law, tectonic properties and evolution of different lithological units in pore water in three dimensions and is directly controlled by landform, hydrological network, sedimentary environment, tectonic properties and evolution history. In areas with complex hydrogeological conditions, the groundwater system often has heterogeneity and anisotropy, and the hydraulic gradient and flow network shape in various parts of the hydrodynamic field will also change with time, so the groundwater hydrodynamic field has temporal and spatial variability.

It is of great significance to correctly establish the hydrogeological conceptual model for groundwater resources evaluation, especially in areas with complex hydrogeological conditions, how to correctly generalize the hydrogeological conceptual model according to exploration data is the key to carry out groundwater system observation and prediction. For example, in the Yangtze River Delta region, the Quaternary sedimentary layer is thick and complex in structure, and the hydraulic connection between aquifers is extremely complex. Improper generalization of hydrogeological conceptual models will easily lead to double or deficient counting of groundwater resources.

Flow field analysis is an important means of hydrodynamic field research, it uses the spatial and temporal properties of potential distribution systems and hydrodynamic principles to reveal the mechanisms of water distribution and exchange in groundwater systems (Zhang *et al.*, 2007). Flow field analysis can determine the non-homogeneity and anisotropy characteristics of the aqueous medium and the location of strong and weak run-off zones, the hydraulic property and exchange capacity of the boundary, and the conditions of groundwater recharge, storage, runoff and discharge, and its main recharge direction, location of the supply source.

Groundwater flow field analysis includes natural flow field analysis based on long-term observation and various artificial flow field analysis based on large-scale pumping (discharging) tests. Through the research of various normal and transient information in flow field analyses, especially the key research on transient information, the spatio-temporal variability of the hydrodynamic field can be fully grasped. Transient information includes excitation information in various states, which can be obtained purposefully and controllably by artificial flow field analysis. For example, the relationship between the structural parameters of groundwater system (conductivity coefficient and storage coefficient) and hydraulic gradient, drop of water level, and conduction rate can be analyzed through the process of constant flow pumping, to further judge the anisotropy and heterogeneity of groundwater system and determine the location of strong and weak runoff zones. The relationship between boundary flow and water level rise etc., can be analyzed by water level recovery process in pumping (discharging) test, to further judge the water supply (discharge) capacity of different boundaries, then determine the main recharge direction and location of the main recharge source, and the water distribution and exchange intensity of the whole system.

Taking Songliao Basin as an example, the temporal and spatial variability of groundwater dynamic fields is briefly described. The hydrodynamic field in Songliao Basin has obvious asymmetry, the edge of the basin, mainly in the north and east of the basin, develops relief-driven centripetal flow (Figure 3.2). The central depression area develops centrifugal flow and cross-formational flow. The southern part of the basin is characterized by cross-formational flow with evaporation. Moreover, the hydrodynamic field in Songliao Basin has vertically zoning. In the centrifugal flow area, three zones can be divided including strong centrifugal flow zone, weak centrifugal zone, and stagnant zone. In the relief-driven centripetal flow region, three zones can also be divided as follows: strong strong relief-driven centripetal flow zone, weak relief-driven centripetal flow zone, and stagnant zone (Lou *et al.*, 2006).

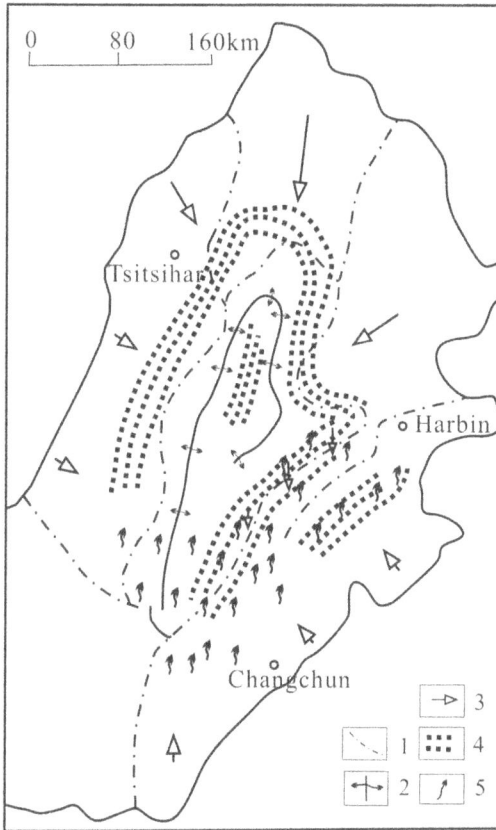

Figure 3.2. Distribution of present hydrodynamic field of Songliao Basin (Lou *et al.*, 2006): 1: first-order tectonic boundary; 2: centrifugal flow at sedimentation and burial stage; 3: relief-driven centripetal flow at elevated and denudated stage; 4: discharge by cross-formational flow; 5: concentration by evaporation.

3.1.2 *Multi-process coupling of hydrochemical field*

The study of the hydrogeochemical process of groundwater systems is helpful to understand the influence of water–rock interaction and human activities on groundwater chemistry, further explore the sources of pollutants, and analyze the causes of temporal and spatial changes of groundwater chemistry. Understanding the evolution law of the hydrochemical field, especially the multi-process coupling in the field, is crucial to the sustainable development and effective management of groundwater resources (Tiwari *et al.*, 2016).

The main ion in groundwater can provide hydrogeochemical information, thus estimating the process and evolution characteristics of the hydrochemical field. At present, a variety of hydrochemical analysis methods and multivariate statistical analysis methods are widely used to analyze hydrogeochemical processes at regional and watershed scales (Cortes *et al.*, 2016; Dehbandi *et al.*, 2017; Taufiq *et al.*, 2018). Hydrochemical analysis methods mainly include Piper diagram analysis, boxplot analysis, and correlation analysis, which are used for hydrochemical composition and source analysis. Multivariate statistical analysis includes principal component analysis (PCA), cluster analysis (CA), and factor analysis (FA). It is mainly used to identify hydrogeochemical processes.

The uncertainty of determination and calculation of groundwater chemical composition directly affects the uncertainty of hydrogeochemical process analysis. Since groundwater is a multi-component solution, the behavior of ions or molecules is somewhat different from that of ideal solution (the size and shape of various molecules are similar, and the potential energy of different kinds of molecules is similar). The interaction (mutual collision and electrostatic attraction) between ions or molecules in water slows down the chemical reaction rate relatively, and some ions do not participate in the reaction. Therefore, if the measured concentrations of the components in water were applied in the calculation without preprocessing, it would result in deviation in some degree. Generally, the measured concentrations should be corrected and using effective concentration (activity) in the calculation to ensure the accuracy.

Major geochemical processes in water–rock interactions include dissolution-precipitation, adsorption-desorption, oxidation-reduction and mixing. The uncertainty of them is shown as follow.

1. *Dissolution–precipitation*: Dissolution–precipitation in the water–rock interaction is controlled by carbonate balance, dissolution, and precipitation.

(1) Carbonate equilibrium: The carbon cycle is one of the most important material cycles in the earth system. Carbonate minerals (calcite and dolomite) are one of the most widely distributed minerals in the solid earth. Carbonate–water interaction plays an important role in controlling the formation and evolution of groundwater chemical composition. The carbonate component in groundwater, CO_2 in the

atmosphere, and the carbonates in rocks together form a carbonate equilibrium system, and the chemical reactions between them play an important role in the formation and evolution of the chemical composition of groundwater.

According to modern hydrochemical theory, carbonic acid in water exists in three combined forms, namely:

(i) Free carbonate, including $CO_2(aq)$, and H_2CO_3, is customary written as "H_2CO_3".

(ii) Bicarbonate, HCO_3^-, is one of the main anions in natural water.

(iii) Carbonate, CO_3^{2-}.

Current hydrochemical calculations for carbonate equilibrium are based on the proportion of carbonate distribution in groundwater at $25°C$. Calculations of three carbonates in groundwater other than $25°C$, particularly at high temperatures, are rarely involved. However, the groundwater temperature varies with seasons and depths, and there are bound to be some errors in calculating groundwater chemical parameters only based on carbonic acid balance at normal temperature. Table 3.2 shows the relative proportion of three forms carbonate in water at $1-100°C$. Although the pH value and the distribution of three carbonates vary small with temperature, it also creates some uncertainty in water chemistry analysis.

(2) Dissolution and precipitation: Dissolution is one of the important roles in controlling the formation and evolution of groundwater chemical composition in water–rock interactions, which can be divided into complete and incongruent dissolution. As a widespread mineral in nature, calcite's dissolution in groundwater is closely linked to the carbonate equilibrium process. Therefore, in the research area where calcite is widely distributed, it is necessary to consider the influence of the uncertainty of carbonic acid equilibrium with temperature on dissolution.

When judging the dissolution of minerals, the influence of the common ion effect and salt effect on the dissolution balance should be considered. The common ion effect refers to if there's the same ions as the mineral dissolution in the solution, the solubility of the mineral will decrease. The salt effect refers to the phenomenon where the solubility of the mineral is lower in pure water than that in highly salty water. When the common ion effect and salt effect exist alone, the

Table 3.2.   Relative proportion of three carbonic acids in water at 0–100°C (Yan *et al.*, 2011).

Temperature (°C)	pH	The proportion of three carbonic acids (%)		
		Free carbonic acid	Bicarbonate	Carbonate
0	8.60	0.9343	98.1468	0.9189
5	8.54	0.9291	98.1217	0.9492
10	8.48	0.9439	98.0928	0.9632
15	8.42	0.9779	98.0603	0.9618
20	8.38	0.9846	98.0252	0.9902
25	8.34	1.0071	97.9877	1.0052
30	8.31	1.0212	97.9482	1.0306
35	8.28	1.0496	97.9069	1.0435
40	8.26	1.0667	97.8645	1.0689
45	8.24	1.0961	97.8209	1.0830
50	8.23	1.1118	97.7769	1.1113
55	8.22	1.1382	97.7321	1.1296
60	8.22	1.1485	97.6873	1.1642
65	8.22	1.1680	97.6423	1.1897
70	8.22	1.1963	97.5977	1.2060
75	8.22	1.2338	97.5530	1.2132
80	8.23	1.2510	97.5090	1.2401
85	8.24	1.2760	97.4652	1.2588
90	8.26	1.2787	97.4222	1.2991
95	8.27	1.3176	97.3806	1.3019
100	8.29	1.3332	97.3392	1.3277

concentration of each solute in the solution has a relatively obvious change trend. When the common ion effect and the salt effect exist together, the common ion effect is greater than the salt effect. Scholars often consider the common ion effect and ignore the calculation of the salt effect, this would bring uncertainty in reaction process analysis in hydrochemical field.

Precipitation effects include saturation precipitation, incongruent dissolution, and decarbonization. The saturation index is commonly used to judge the saturation state of groundwater relative to minerals, and precipitation occurs when minerals in the groundwater reach a saturated (or supersaturated) state. Incongruent dissolution means that when some complex minerals are dissolved, their products often generate new solid components (minerals) in addition to dissolved components, thus generating precipitates. Such as albite and potassium, when they are dissolved, secondary solid mineral kaolinite has generated beside releasing Na^+ and K^+ into the solution. When multiple minerals are present in the aquifer, although the dissolution of a single mineral may be congruent dissolution, the dissolution of one mineral would cause precipitation of another mineral, because of various solubility for minerals, which brings the challenge for hydrogeochemical study in areas with complex hydrogeological conditions, and increases the uncertainty in the study of water–rock process. In the hydrochemical field, CO_2 escape caused by the change of temperature and pressure, and the reduction of the content of HCO_3^-, Ca^{2+}, Mg^{2+}, and Fe^{2+} in water is called decarbonization. Through decarbonization, CO_2 in water escapes, while HCO_3^- with Ca^{2+}, Mg^{2+}, and Fe^{2+} forms precipitation (take Ca^{2+} for example).

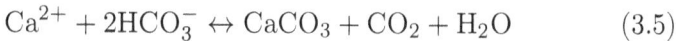

$$Ca^{2+} + 2HCO_3^- \leftrightarrow CaCO_3 + CO_2 + H_2O \qquad (3.5)$$

Under the combined action of carbonic acid balance, salt effect, common ion effect, incongruent dissolution, and decarbonization in areas with complex hydrogeology conditions, the uncertainty of hydrogeochemical process analysis will be greatly increased, and various chemical analysis methods are needed to reduce the uncertainty of chemical reaction process determination.

2. *Adsorption and desorption*: The adsorption of ions refers to the enrichment of solutes (ions or molecules) in solution at the interface between solid and liquid and also refers to the difference of ion concentration in the solution and the diffusion layer near the solid and liquid interface. The opposite process of adsorption is desorption, and all factors affecting adsorption will also affect desorption.

Adsorption isotherm refers to the relationship curve between the equilibrium concentration of adsorbents in solution and the content of adsorbents on the surface of solid particles under constant temperature conditions, which is often used to express the degree of water–rock adsorption. The commonly used isothermal curve equations include Langmuir, Freundlich, Tamkin empirical equations.

The factors that affect the adsorption process include the properties of solid and liquid medium and external factors, which are described as follows:

(1) Properties of solid and liquid medium: The texture, clay content, colloid quantity, ionic charge quantity, and material composition of solid media will affect the adsorption process and adsorption capacity (Huang *et al.*, 2016; Nielsen, 1972; Wang *et al.*, 2005). The solute in the liquid also affects the adsorption. For example, high concentrations of chloride and potassium sulfate in soil have inhibitory effects on the adsorption of nitrate nitrogen in soil (Heilman, 1975). The increase of organic matter content in the solid medium will also influence the adsorption process. Liu *et al.* (2005) found that organic matter is the main factor controlling the adsorption of ammonium nitrogen in the soil.

(2) External influencing factors: The change of external temperature has a great influence on adsorption and desorption. For example, temperature affects the adsorption and desorption of shale. With the increase of temperature, the adsorption ability of shale decreases (Guo *et al.*, 2013). Wang *et al.* (2012) found that freezing and thawing can promote the adsorption of exogenous cadmium in soil, and the adsorption amount, and adsorption rate of cadmium increase with the freezing and thawing times.

3. *Oxidation-reduction*: In hydrogeochemical research, redox conditions directly or indirectly control the migration characteristics of many elements. The redox state in the groundwater system mainly depends on the amount of oxygen circulating into the system, the amount of oxygen consumed by bacteria decomposing organic matter, or oxidizing low valence metal sulfides, iron-containing silicates, and carbonates (Xing, 2012). If the amount of oxygen entering the system is greater than the amount of oxygen consumed, the system is in an oxidized state, otherwise, it is in a reduced state. Redox potential (Eh) is an important parameter to judge the redox conditions of groundwater. The Eh of shallow groundwater is generally positive showing an oxidized state, while Eh is negative, groundwater is in a reducing state. Many factors are affecting the redox conditions of the groundwater system, mainly including natural factors and human factors that can be described as follows:

(1) Natural factors: These include the type and quantity of oxidizing agents and reducing agents in groundwater, the recycling process of groundwater, and the type and quantity of microbes and

organic matter in groundwater (Money Club, 2012). According to the order of decreasing oxidation capacity and reduction capacity, the common oxidants in groundwater are mainly O_2, NO_3^-, NO_2^-, Fe^{3+}, SO_4^{2-}, S, CO_3^{2-}, and HCO_3^-, the common reducing agents are mainly organic matter, H_2S, FeS, NH_4^+, and NO_2^-. Each oxidant and its corresponding reducing agent form a single redox system. When the composition of oxidant and reducing agents in the hydrochemical field is complex, the redox potential of groundwater is often between the redox potentials of a single system, which is close to the redox potential of a single system with large content. The circulation process of groundwater will affect the temporal and spatial distribution of redox potential in the system. In the recharge area, groundwater is often in a strong oxidation state due to sufficient dissolved oxygen in the water. With the runoff of groundwater, dissolved oxygen in the water is continuously consumed, and NO_3^- is converted into an oxidant to be preferentially utilized and reduced to NO_2^- and N_2 by denitrification. With constantly consuming of NO_3^-, the groundwater system transforms to a suboxidation state in which Fe^{3+} and SO_4^{2-} are used for organic matter oxidation in the aquifer. When SO_4^{2-} is also depleted, the aquifer is in a state of reduction. Microorganism in groundwater oxidize organic matter using oxidants such as O_2 and NO_3^-, so that the system is gradually reduced.

In the groundwater system with complicated hydrogeological conditions, hydrochemical fields usually contain multiple single redox systems, so it is difficult to analyze the redox process based on Eh and solute concentration. Also, due to the long runoff path of groundwater, there are great differences in redox conditions and processes from recharge area to discharge area and from shallow to deep groundwater. Typical monitoring boreholes need to be arranged to realize long-term and regular groundwater chemical monitoring.

(2) Human factors: Organic pollution, mining and groundwater exploration etc., can change the redox conditions of groundwater to varying degrees, resulting in changes in the chemical components. A series of redox reactions occur when organic pollutants are released into the aquifer, and the degradation of organic matter is the oxygen loss process of the hydrochemical field. Artificial mining and pumping and drainage processes would change the redox conditions of groundwater, accelerate circulation of groundwater, and allow more

dissolved oxygen to enter the groundwater, resulting in an oxidized condition. Under oxidation conditions, heavy metals are likely to react with oxidants and enter into groundwater with strong migration ability, posing a threat to ecological safety and human health (Wang, 2005).

4. *Mixing*: When different kinds of natural groundwater are mixed, mixed dissolution or precipitation will occur. Generally, only when the components of the two aqueous solutions and their mixing ratios are properly matched can the mixed dissolution occur in the strict sense, otherwise precipitation will accompany it (Qian and Hu, 1996).

The following is an example of a brief description of the occurrence of mixed dissolution and precipitation.

(1) When HCO_3^- and Ca^{2+} concentration in one aqueous solution are higher than that in the other, the mixed dissolution often occurs.
(2) When the concentration of HCO_3^- in a solution is higher than that in another solution, while Ca^{2+} concentration is lower than that in another solution, the mixed precipitation often occurs.
(3) When the concentrations of HCO_3^- and Ca^{2+} in the two aqueous solutions differ greatly, and the pH value of the aqueous solution with a higher concentration of HCO_3^- is lower mixed dissolution and mixed precipitation may occur when the two aqueous solutions are mixed in a certain proportion (Qian *et al.*, 2007).

Therefore, due to the uncertainty of groundwater composition and the mixing quantity and spatial location, the determination of the mixing process and degree is complicated. At present, there are few research on the uncertainty of groundwater mixing, and more comprehensive researches need to be performed.

The uncertainties and influencing factors for the major hydrogeochemical processes in aquifer hydrochemical fields are briefly described above. The hydrogeological conditions of groundwater aquifer are complex, and various hydrogeochemical processes often occur in the coupling, which greatly increases the uncertainty and difficulty of hydrochemical field observation and prediction. Also, the hydrochemical characteristics are continuously changing under the influence of natural and human factors, so it is necessary to regularly monitor the hydrochemical field in the study area to weaken the uncertainty (Zhang *et al.*, 2019).

3.2   Observation Scale and Uncertainty

3.2.1   *Up-down scaling and scale conversion*

The scale problem is a common problem faced by many disciplines. From the perspective of hydrogeological structure, aquifers have multi-scale characteristics, and different scales are selected according to different research purposes (Qin *et al.*, 2014). At present, there are two popular classification methods in the world. The first is proposed by Dagan (Dagan, 1986) to divide aquifer into three basic scales, laboratory scale, local scale, and regional scale. The second is proposed by Koltermann and other (Koltermann and Gorelick, 1996) that classifies aquifer scale into multiple pore-scale, flow characteristic scale, stratiform characteristic scale, flow channel scale, sedimentary environment scale, and watershed scale etc., according to the sedimentary structure, sedimentary environment, and geological characteristics of pore medium. The main characteristics of each scale are shown in Table 3.3.

The scale problem is important in the numerical simulation of groundwater flow and mass transport. The scale of data collection is different from (usually smaller than) the scale used for discrete aquifers in numerical models. For example, the hydraulic conductivity of field samples (e.g., core measurements, slug tests, or packer tests) has measurement support of centimeters to meters, while the numerical model of groundwater flow requires a conductivity representative of tens to hundreds of meters (Wen and Gómez-Hernández, 1996). Scale conversion (upscaling) has become one of the difficult and hot issues in groundwater seepage and solute transport research in recent years. Upscaling methods usually include a statistical average method, volume average method, homogenization theory, and renormalization technology, and upscaling usually includes parameter upscaling and process upscaling. As upscaling is very important to solve the problem of pollutant migration in the field, it has received extensive attention from environmentalist and hydrogeologists in the past 20 years. At present, the research on parameter upscaling mainly includes permeability upscaling of heterogeneous aquifer, fissure network model, the influence of scale-up on solute transport, and upscaling of seepage and solute transport in highly heterogeneous three-dimensional aquifers. Frippiat and others have compared two upscaling methods for solute transport in heterogeneous

Table 3.3. Scale classification of sedimentary heterogeneity.

Scale	Scaling factors of similar length/m	Geological characteristics	Factors affecting heterogeneity	Observation/ Measurement technology	Volume average of hydraulic conductivity observations
Multiple observation scales	10^{-6}	Particle size and shape, sorting, packing process, component cementation, and clay fissures	Source of pore structure, diagenesis, sedimentary process, and sediment migration mechanism	Observe the cross-section of rock samples, the structure of cuttings aggregates, and collect lens bodies	Involving several pore structures (observed by micropermeameter)
Flow feature scale	10^{-3}	Basic sedimentary structure, small water flow, cross-penetrating strata, bedding structure, dislocation of fracture distribution, deformation of soft sedimentary layer, etc.	Inhomogeneous diagenetic process, sediment transport mechanism, biological activities	Core sampling, manual sampling, and outcrop observation	Core sampling (measured by permeameter)
Stratigraphic characteristic scale	10^{1}	With rich sedimentary structure, the subsidence classification process	Layer boundary, fine passage, belt dune	Observation of outcrops, lithology, and analysis of geophysical logging	The influence range of the well

Flow channel scale	$10^1 - 10^2$	Geometric structure, bedding type, and ductility, lithology, and mineral content of flow channels	Frequency of shale formation, geometric structure of sand layer and shale body, sedimentary filling combination	Observation of outcrops, well profile X-ray fault images, lithology, and geophysical logging	Local test (short-term pumping or tracing test)
Sedimentary environmental scale	$10^2 - 10^4$	Multiple composite phase relation and formation of morphological characteristics	Control of flow and the sedimentary mechanism by fissures and internal basins	Map, stratigraphic profile lithology and geophysical logging, seismic profile logging	Regional experiment (long-term pumping or tracer test)
Watershed scale	Greater than 10^5	Geometric characteristics of the watershed, geometric characteristics and structural characteristics of strata, lithological discontinuity and regional strike of strata	Faults (sealing), folded crustal structure, the influence of sea level and climate changes, stratigraphic thickness, strike, unconformity	Maps, seismic test sections, and cross-sections	Shallow crust properties

aquifers: the upscaling of heterogeneous parameters (dispersion) and the upscaling of transport equations (essentially the same as the above parameter upscaling and process upscaling). Each method includes three models. Dagan and others systematically introduced the scale-up models and methods for seepage calculation in different heterogeneous media, including layered aquifers, three-dimensional anisotropic aquifers, aquifers in stationary and non-stationary random media, and the scale-up of seepage near wells.

At present, scale conversion in hydrogeological research has mainly solved two problems: one is the description of heterogeneity and randomness of aquifers at different scales. The Monte Carlo method and geostatistics (Kriging) method can be used to quantitatively describe the spatial random distribution of aquifer permeability coefficient, thus studying the seepage and solute transport problems of heterogeneous aquifers; the second is to study the seepage and solute transport in heterogeneous aquifers by stochastic numerical simulation technology. Through the numerical simulation of seepage and solute transport in large-scale aquifers, the temporal and spatial distribution of groundwater variables in large areas can be calculated, and the boundary values of small-scale domains embedded in large-scale can be obtained. The seepage and solute transport problems in the small-scale domain are calculated, and then the calculation is carried out in the large-scale. This method is a numerical analysis method of upscaling and downscaling. Because scale transformation involves the scaling of a probability distribution, the direct upscaling method is called the random method (Bierkens and van der Gaast, 1998).

To describe the uncertainty of groundwater seepage and solute transport, people use random methods to describe the heterogeneity of aquifer media and use stochastic partial differential equations to describe seepage and solute transport. As a result, solving stochastic partial differential equations becomes the key to solving uncertain.

Groundwater stochastic simulation methods mainly include the moment equation method, Kriging interpolation method, and Monte Carlo method. The moment equation method obtains the solution of the stochastic problem by solving the stochastic partial differential equations about the mean and covariance, while the Monte Carlo method is a computer simulation method to simulate the stochastic process by averaging a series of deterministic problems reflecting the actual properties of the aquifer (Wang and Fan, 2017).

The Monte Carlo method is the most commonly used method to solve stochastic partial differential equations through numerical simulation. The Monte Carlo method, also known as statistical simulation method and random sampling technology, is a very important numerical calculation method based on "random number" and probability statistics theory, which has been widely used. The Monte Carlo method is based on a computer with dense information and high-speed calculation, through scientific and reasonable statistical modeling, complex research objects or calculation problems are transformed into simulation and calculation of random numbers and their digital characteristics, thus essentially simplifying the research problems, reducing the calculation complexity, and obtaining approximate solutions with excellent properties (Xue and Wu, 1999).

It always could find ways to solve various problems based on Monte Carlo method. Overall, based on the association of the thing itself with the stochastic process, the practical solution using Monte Carlo is roughly classified as two types.

The first is deterministic mathematical problem. When Monte Carlo method is adopted to deal with such questions, firstly, an associated probability model is constructed according to the problem itself, and the solution of deterministic problem is exactly equal to the probability distribution or the mathematical expectation of the model constructed. Next, a large number of random experiments are started on this probability model, that is, a large number of random numbers and random variables are generated. Finally, the sampling arithmetic average value of the variables determined by the model is approximately equal to the estimated value of the deterministic problem. Calculating inverse matrices, multiple integrals, solving Riemann integrals, algebraic equations, boundary value problems of individual partial differential equations, and calculating eigenvalues of differential operators all belong to this type.

The second is a random problem, which can be seen everywhere in life. For example, the diffusion of neutrons in the medium. This is because neutrons in the medium are not only affected by some deterministic factors but also by some uncertain and random factors. For such issues, although it can be expressed as some special functional equations or specific multiple integrals, and further converted into a random sampling method to calculate, this indirect simulation method is not usually selected, but adopts a direct simulation method, that is, according to the probability rules of actual physical characteristics, scientific computers are used to carry out simulated random sampling tests. Nuclear physics problems, inventory problems in operations research, random queuing problems in service systems, the ecological competition of organisms, and the spread of infectious diseases all belong to this type.

Using the Monte Carlo method to simulate groundwater flow and solute transport has the advantages of easy understanding and operation and has been accepted by more and more researchers. Chen and Wu (2005) used the Monte Carlo method to study the influence of spatial variability of aquifer permeability coefficient on groundwater numerical simulation results. Yan and Wu (2006) also studied the influence of the mean and variance of the permeability coefficient on the two-dimensional spatial distribution of pollution plumes by the Monte Carlo method. Although the aquifers involved in the above study have spatial variability, they are relatively uniform (the variance of lnK is generally less than 0.5, and the permeability coefficient conforms to the assumption of lognormal distribution). The heterogeneity of the actual aquifer in the field is very complicated, and the zoning treatment of heterogeneity is generally inevitable. Also, it is very common to sandwich large sub-sandy or sub-clay lens in the sand layer. When the research scale is not very large, it cannot be simply treated by statistical average and must be zoning treated according to heterogeneity. When studying the influence of aquifer heterogeneity zoning on the Monte Carlo simulation results of groundwater, two methods of corresponding filling and arrangement filling are adopted to generate the final random field of heterogeneity zoning of permeability coefficient under different realistic numbers, and the Monte Carlo simulation results of the two methods are compared.

Due to the limitation of the groundwater environment, water samples are collected from discrete wells (sample points). Spatial interpolation is used to deduce the distribution map of the whole area from the limited discrete water sample data. Using the spatial interpolation method, groundwater quality evaluation has changed from point analysis to regional spatio-temporal evolution analysis, realizing the process from point to surface and improving the scientificity and effectiveness of groundwater quality evolution laws (Bai and Yin, 2010). The Kriging method is a very useful and flexible geostatistical gridding method, which can investigate the spatial autocorrelation of images (Zhao, 2010). The Kriging interpolation method is also called the spatial local interpolation method, it is based on variogram theory and structural analysis, to realize unbiased optimal estimation of regionalized variables in a limited area, it is one of the main contents of geostatistics. It was first proposed by French geographer Matheron and South African mine engineer Krige and applied to mine exploration. Its principle is to assume that the spatial change of a certain attribute is neither completely random nor completely determined (Qing *et al.*, 2010). The core of Kriging interpolation is to create a semi-variogram. Semi-variogram can describe not only the spatial structural changes but also the random changes of regionalized variables (Ling *et al.*, 2012).

At present, there are the following problems in the research of groundwater scale conversion: first, how to analyze seepage and solute transport in heterogeneous aquifers at different scales in the application, especially the problem of pollutant multiphase flow transport; second, the scale conversion formula for groundwater calculation in heterogeneous aquifers has not been given theoretically; third, how to analyze the heterogeneous characteristics of aquifers of different scales with limited hydrogeological monitoring data, and how much monitoring data are needed to analyze the heterogeneous aquifer characteristics of different scales (Qin *et al.*, 2014).

3.2.2   *Application scope and uncertainty analysis of the scale conversion model*

The regional groundwater system is a complex open system, influenced by factors such as hydrometeorological conditions, geological structures, topographic features, vegetation, and human activities (Wu and Zeng, 2013). A large number of field tests have well confirmed that solute transport in heterogeneous aquifers can be described by stochastic theoretical models. The research on physical heterogeneity is more in-depth, while the research on chemical heterogeneity is relatively weak. From the influence of the size of the test site and the test duration on the prediction accuracy of the model, the stochastic theoretical model has scale dependence, which is the scale problem mentioned in the previous section. When these methods are used for scale conversion, a series of uncertainties may be brought. Moreover, the uncertainty of aquifer heterogeneity also affects the model prediction.

The moment equation method primarily controls statistical moments (usually pre-second-order moments) of random variables by derivation of a series of deterministic partial differential equations or some closed approximation. For large-scale problems, the disadvantage of the method is that the calculation is too large.

The Monte Carlo method is a method of statistical sampling, whose accuracy depends on the number of Monte Carlo simulations when generating a random field. However, the direct sampling Monte Carlo method is simple in concept and easy to implement, but its disadvantage is that the computation is too large, especially for large-scale problems. In Monte Carlo simulations, fewer physical nodes would lead to unstable results. In direct sampling Monte Carlo simulations, a large number of implementations are required to get statistically accurate results (Li and Zhang, 2007).

The Kriging interpolation is suitable for that there is a spatial correlation of regionalized variables, i.e., if the various function and structural analysis indicate a spatial correlation of regionalized variables, the Kriging interpolation can be used for interpolation or extrapolation. The essence of this is using the raw data of regionalized variables and the structural characteristics of the various functions, to make linear unbiased, optimal estimates of unknown sample points. Unbiased refers to the mathematical expectation of a deviation is 0, and optimal is the quadratic sum of estimated and real values is minimum. Therefore, the Kriging interpolation method is a linear unbiased optimal estimation of unknown sample points based on some known sample data in the finite field. When describing the spatial variability of aquifer parameters based on stochastic theory, the traditional geostatistical simulation methods often assume that aquifer structural parameters (such as mean, variance, and covariance) are known, but these parameters are unknown for real field investigation.

3.3   Simulation Uncertainty

In recent years, there have been many researches on uncertainty analysis of groundwater simulation. To obtain simplified and reasonable modeling uncertainty statistics and meet the actual needs of hydrogeologists, more and more researchers (Xu *et al.*, 2006;

Ajami *et al.*, 2007; Blasone *et al.*, 2008; Hassan *et al.*, 2008; Rojas *et al.*, 2008; Jin *et al.*, 2010; Renard *et al.*, 2010; Wu *et al.*, 2011; Zheng *et al.*, 2013) have accepted the method of using the framework to analyze groundwater model uncertainty. Generally speaking, according to the logical process of groundwater simulation, the uncertainty comes from three sources: model parameters, conceptual model (or model structure), and observation data. About the effect of the three sources on model uncertainty, some scholars consider the uncertainty of parameters > uncertainty of the model > uncertainty of the data, although it is accepted that 5–30% of the uncertainty comes from the model (Rojas *et al.*, 2008). Even though, the focus on conceptual model is still insufficient.

There are many classifications of simulation uncertainty sources. For example, (1) natural uncertainty caused by the inherent randomness in natural processes; (2) uncertainty comes from the deficiencies of the model and cannot represent real physical processes; (3) uncertainty of model parameters; (4) uncertainties caused by observational errors; (5) operational uncertainty caused by human factors (Li *et al.*, 2009). According to the nature of the subject, the source can be interpreted as random, fuzzy, grey, and unknown uncertainties (Liu and Shu, 2008). Also, according to the causes, the sources can be divided into uncertainties caused by natural processes, human activities, and cognitive abilities. Merz and Thieken classify uncertainties into two categories: accidental uncertainties and cognitive uncertainties (Merz and Thieken, 2009). Also, these studies summarize and discuss the uncertainties of groundwater numerical simulation from different points (Katz *et al.*, 2002; Helton and Oberkampf, 2004; Singh *et al.*, 2010; Yang *et al.*, 2010).

The core issue of current numerical simulation is to prevent simulation distortion and strive to improve the simulation. There are many kinds of simulation distortion, the most important of which is the distortion of a conceptual model. Once the hydrogeological conceptual model cannot describe the basic movement law of groundwater in the study area, no matter how fine the subsequent work is, it cannot make up for the mistake of establishing the conceptual model. However, this problem has not attracted the general attention of hydrogeological numerical simulators. Without careful analysis of hydrogeological conditions and basic conditions of groundwater flow, a conceptual model is hastily given. The prediction results of this model are incredible. Although the fitting degree of some models in a short period is not very poor, the fitting degree is not the only criterion to test the models. The most important and basic thing should be that the models should conform to the basic hydrological characteristics of the groundwater system and the simulation method should conform to the mechanism. However, some studies often make a brief mention of the process and basis of the establishment of the conceptual model, citing other people's model structures without thinking about them, which makes the conceptual model have many defects and lack credibility (Chen, 2003).

Due to the complexity of the groundwater system, we can't fully understand the system, including the thickness of geological units, the characteristic of aquifer (location, type, etc.), boundary conditions, the temporal and spatial variability of hydrogeological conditions, etc. The understanding of groundwater systems comes from limited hydrogeological survey data, which is far from fully reveal the structure of the groundwater system, especially the serious clay–sand mixed layer in the Quaternary aquifer, which makes the model stratification extremely difficult (Wu *et al.*, 2003; Zhang *et al.*, 2004; Li *et al.*, 2005; Hu and Chen, 2006a, 2006b; Hu *et al.*, 2007).

Modeling is an important means of groundwater system research. Conceptual models or numerical models based on field observation, indoor experiments, and mathematical equations make groundwater system research convenient and efficient. At the same time, they also provide a reliable basis for groundwater resources planning, exploitation, protection, and evaluation (Hu *et al.*, 2011). The application of the conceptual model and mathematical model in environment, geology, and other aspects has effectively promoted the research of groundwater science. However, when the conceptual model and mathematical model are applied to the actual observation and prediction of the groundwater system, there is often a significant difference between the predicted value and the actual observation value.

The uncertainty of parameters and scales has been discussed in the previous section. The uncertainty caused by the parameters and scales will be transferred to the model, which will affect the accuracy of the model and bring uncertainty. However, the main uncertainty of the modeling does not come from the parameters and scales, the defects of the conceptual model and the solution error of the mathematical model are more critical (Figure 3.3).

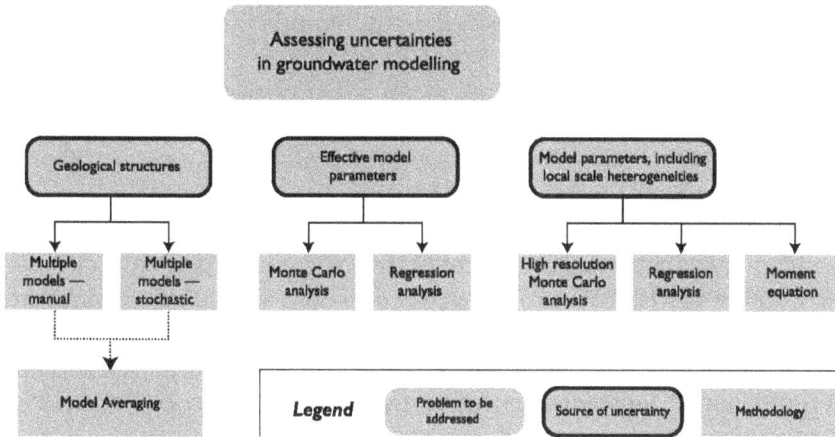

Figure 3.3.   Classification of existing strategies for the assessment of geologically related uncertainties in groundwater models (Refsgaard *et al.*, 2012).

When converting the actual groundwater system into a conceptual model, the following steps need to be handled according to the general process:

(1) Dividing the boundary of the conceptual model.
(2) Defining the type of boundary.
(3) Defining the number, sequence, and type of aquifers and aquitards.
(4) Defining the types of hydraulic connections and material transfer between aquifers.

The conceptual model can qualitatively give what hydraulic connections and material transfer processes occur in the groundwater system in the study area and often only distinguish the strength of hydraulic connections and material transfer, making a qualitative "portrait" of the groundwater system.

The geological model is one of the elements of a conceptual model. Conceptual models of groundwater systems may differ in different respects, such as (a) alternative geological structures interpreted by different geologists (Hojberg and Refsgaard, 2005; Troldborg *et al.*, 2007); (b) alternative concepts of flow and transport processes, such as porous media, discrete fissure networks, or river channel networks (Selroos *et al.*, 2002); (c) partition and characteristics of the optional hydraulic parameter (Poeter and Anderson, 2005); (d) aggregation of different model layers and different boundary conditions (Rojas *et al.*, 2008, 2010).

The geological model is based on existing data and knowledge, that is to say, it is a simplification of unknown natural systems (reality). Due to the incompleteness of real information, it is possible to construct several reasonable geological models, that is, based on existing data and knowledge, they cannot be excluded. In the multi-modeling method advocated by Neuman (2003) and Poeter and Anderson (2005), groundwater flow and migration models are established for multiple alternative geological models of the same site. A recognized problem with this method is that the selected conceptual model will never cover the space of all reasonable models. The shortage of conceptual models is partly due to practical reasons, the number of models usually needs to be limited, partly due to geologists' insufficient imagination of other reasonable models, and finally conceptual models can not explain unknown information (uncertainty caused by ignorance). This kind of incomplete coverage of model space usually leads to conceptual uncertainty and prediction uncertainty.

There are two main ways to build multiple models:

1. *Multiple geological models — Manual*: This is currently the most commonly used method of multiple modeling. Geological models are established by manual interpretation.

2. *Multiple geological models — Stochastic*: The program focuses on the random generation of geological models and can use alternative concepts such as flow and transport equations and alternative boundary conditions. If the assumption is stationary, random generation can be used in the whole geological field. Alternatively, it may supplement artificial geological interpretation within a sub-area and/or one or more geological structural units.

For stochastic geological tools such as TProGS (Carle and Fogg, 1996), geological data are analyzed using geostatistical methods, for example, through semi-variograms and transformation probabilities to describe geological heterogeneity. Another technique is based on multi-point geostatistics (Luo *et al.*, 2015), in which training images are used. Both techniques allow the combination of subjective geological knowledge and geostatistical analysis. On this basis, some reasonable geological models can be generated. Stochastic methods are often used to fill some structural heterogeneity, whose scale is smaller than that of whole geological structural units. For example, TProGS is used to generate a geological reality that the artificially interpolated geological unit "moraine layer" can be decomposed into smaller clay and sand sequences (Troldborg *et al.*, 2007). In this way, the stochastic method can be used to study the influence of local scale inhomogeneity.

When building a mathematical model based on the conceptual model to solve the problem, the following parameters need to be input:

(1) Elevation and coordinates of the groundwater system.
(2) The governing equation of groundwater system boundary.
(3) Specific structure of groundwater system, including horizon, thickness and location.
(4) Various parameters of groundwater movement and solute transport.
(5) The governing equations of groundwater flow and solute transport.
(6) Initial groundwater system parameters and solute transport parameters.

Then, the simulation results of the mathematical model should be obtained according to the steps of mesh generation, time step setting, and partial differential equation (difference scheme) calculation.

The defects of the conceptual model come from that the division of boundaries, boundary types, aquifer and aquitard structures, hydraulic connections, and material transfer types rely on empirical judgment and limited geological boreholes, and long-term monitoring data. While, in the process of solving the mathematical model, to satisfy the convergence condition, the partial differential equation is solved by the difference scheme, and the nonlinear equation is approximately transformed into the linear equation, which may bring certain errors to the model in the solution process.

3.3.1   *Defects of the concept model*

The following describes the flaws that often occur when conceptual models are established.

The dimension of conceptual model: From the amount of information contained in the model, the three-dimensional model is superior to the quasi- three-dimensional model, the two-dimensional model, and the one-dimensional model. However, the requirement for observation data to establish three-dimensional models is much higher than that for quasi-three-dimensional models and two-dimensional models. At the same time, solving hydrodynamic equations and solute transport equations of three-dimensional models poses higher challenges to programming and computer processing. Based on this, two-dimensional models and quasi-three-dimensional models have been applied more in engineering practice.

The actual groundwater system problem is a three-dimensional geological problem, and the groundwater flow belongs to three-dimensional flow. As the recharge (rainfall, infiltration of canals and irrigation) and discharge (evapotranspiration, springs, and incomplete rivers) all occur on the phreatic surface or its shallow part, and the thickness of the groundwater system could be hundreds of meters, this is the fundamental reason for the formation of three-dimensional flow in the groundwater system. At the same time, the vertical flow of groundwater cannot be ignored in areas with low groundwater buried depth such as riparian zone, coastal area, and alluvial plain area. Therefore, generalizing the three-dimensional groundwater system into a two-dimensional model will bring great errors (Chen and Hu, 2008; Wan *et al.*, 2013; Chen, 2012).

The quasi-three-dimensional model is ignoring the horizontal velocity component in the weak permeable layer. Some scholars pointed out that when the permeability coefficients of the aquifer and the weak permeable layer differ by more than two orders of magnitude, the error caused by using the quasi-three-dimensional model to describe the groundwater system is less than 5%. This case is only applicable to isotropic weak permeable layer; however, there is no complete weak permeable layer in the actual groundwater system, and clay layer and sand layer often appear alternately. It is impossible to completely simulate the interbedded structure of the groundwater system when the conceptual model is established, otherwise, the unacceptable workload will be brought, so the interbedded structure is often treated as a composite weak permeable layer. When the anisotropy ratio of the weak permeable layer is 10, the error caused by generalizing the groundwater system into a quasi-three-dimensional structure has exceeded 27.7%. When the simulation time and the water storage coefficient of the weak permeable layer further change, the error will continue to expand.

Therefore, the groundwater system is generalized into two-dimensional models and quasi-three-dimensional models based on insufficient information that can not to deal with the complex structure of aquifers, the conceptual model would have serious defects. Even if it can fit a small number of observation holes, it can only prove that the location of the observation holes does not reveal the three-dimensional characteristics of the groundwater system and cannot prove the rationality and correctness of the model.

Boundary issues: The division and definition of boundaries is a crucial step in conceptual models. Wrong boundary regions or boundary types will lead to complete distortion or inability to converge in subsequent mathematical model calculations. Choosing a reasonable boundary area and defining a reliable boundary type is the basis for establishing a successful conceptual model. However, the selection and definition of boundaries are not completely correct. In a few cases, artificially divided boundaries bring considerable errors to the construction of conceptual models.

Before delimiting the model boundary, what is the natural boundary? Natural boundary refers to the edge surface of a continuous groundwater body. Therefore, the water-resisting layer contacting groundwater and the contact surface between groundwater and surface water is the natural boundary of groundwater. Also, for the saturated groundwater, the saturated–unsaturated interface is its natural boundary, but it is a movable natural boundary. Although the relationship between surface water and groundwater has different forms and conditions, it is easy to understand. If there is a "water-resisting layer" in the strict sense in nature. Generally, the so-called "water-resisting layer" is relative (Chen *et al.*, 2004, 2006; Hu *et al.*, 2005; Chen *et al.*, 2007; Hu and Chen, 2008).

Subject of groundwater dynamics, hydrogeology, and numerical simulation of groundwater divide the boundaries of groundwater systems into two categories. One is the boundary with known head distribution law and the other is the boundary with known flow change law. When the Dupuit assumption is not allowed to be introduced, the boundary with infiltration recharge, such as phreatic surface, becomes the third boundary, which is with unknown head distribution and flow change law, and needs to be treated by material derivative method or mass conservation method. In practical engineering application, the water boundary between the groundwater system and the surface water body is regarded as the boundary of the groundwater system, of which the variation law of water head can be regarded as the first kind of boundary conditions, and the second kind of boundary conditions generally include the borehole wall of constant flow pumping wells, i.e., constant flow boundary and zero flow boundary such as water-resisting layer boundary and groundwater watershed boundary. When determining the boundary conditions, comprehensive consideration should be given according to hydrogeological conditions and existing comprehensive data (Chen and Wan, 2002; Wang and Chen, 2002; Chen, 2003; Chen *et al.*, 2003, Cheng *et al.*, 2003, Wang *et al.*, 2003).

There is no absolute first-class boundary, second-class boundary, or third-class boundary in the natural groundwater system, nor is there a absolute water-resisting layer. The boundary and water-resisting layer are relative. The boundary conditions are equivalent to known nodes. The richer the boundary conditions, the simpler the conceptual model will be. Under reasonable circumstances, the simplification of boundary conditions will not affect the accuracy of the conceptual model, but the setting of the unnatural boundary (artificial boundary) and the nonholonomic boundary is likely to bring significant defects to the conceptual model, as follows:

The artificial boundary of the first type: The problem of delimiting the first type of boundary through artificial observation hole data lies in the instability of the boundary. The results of groundwater simulation of the first type of boundary thus given are affected by many factors, such as the errors caused by the establishment of groundwater conceptual model and the deviations caused by the approximate solution of groundwater mathematical model. Also, the scarcity of measurement data and observation errors increase the modeling uncertainty. Since the interaction mechanism between various factors is unknown, it is difficult to clearly separate each part of the uncertainty source and describe it independently.

The artificial boundary of the second type: The second type of boundary here usually refers to the zero flow boundary, generally, many unreliable "weak permeable layers" are treated as zero flow boundary, but in fact, the vertical overflow and horizontal velocity in weak permeable layers cannot be ignored. Due to insufficient water quantity and inaccurate water head in monitoring holes of weak permeable layers, it is difficult to find errors in the water head simulation of weak permeable layers. At the same time, the discharge on the groundwater watershed is zero, so it is often treated as a zero discharge boundary in the model treatment. However, with pumping activities and rainfall infiltration, if the movement of the groundwater watershed boundary cannot be reflected in the conceptual model, it will also bring serious calculation errors in the numerical model.

Nonholonomic boundary: When dealing with boundary issues, some boundaries do conform to the characteristics of the first or second boundaries; however, in the natural groundwater system, the boundary is usually nonholonomic. For example, it is often difficult for the surface water system to connect with the whole aquifer group. At this time, when the surface water system is taken as the first type of boundary condition, without special treated for the layered structure, the conceptual model will not reflect the correct groundwater system structure.

3.3.2   *Calculation errors of mathematical models*

When discussing the calculation error of the mathematical model, it is necessary to discuss the definite conditions of the differential equation of groundwater flow firstly. Taking the three-dimensional anisotropic unsteady flow as an example, it generally includes the head function, the first type of boundary conditions, the second type of boundary conditions, and the initial conditions, as follows:

$$\frac{\partial}{\partial x}\left(K_{xx}\frac{\partial H}{\partial x}\right) + \frac{\partial}{\partial y}\left(K_{yy}\frac{\partial H}{\partial z}\right) + \frac{\partial}{\partial z}\left(K_{zz}\frac{\partial H}{\partial z}\right) = \mu_s\frac{\partial H}{\partial t} \qquad (3.6)$$

$$H|_{t=0} = H_0(x, y, z) \qquad (3.7)$$

$$H|_{B1} = H_B(x, y, z, t) \qquad (3.8)$$

$$K\frac{\partial H}{\partial z}\Big|_{B2} = q(x, y, z, t) \qquad (3.9)$$

Equations (3.6), (3.7), (3.8), and (3.9) represent the head function, initial conditions, the first type of boundary conditions, and the second type of boundary conditions. The definitions in the expressions strictly conform to the physical meaning of the corresponding parameters. The K, H, q, and μ_s represent hydraulic conductivity, head, flow, and specific yield, respectively.

In fact, except for a few very simple theoretical conditions (such as isotropy, ignoring the assumption of vertical velocity), an analytical solution can be obtained, in general, a numerical simulation is needed to obtain numerical solutions. Complex boundaries and initial conditions cannot be calculated by continuous differential equations, and finite discrete grids or elements are needed to get numerical solutions. Therefore, two main numerical model calculation methods are generated: finite difference methods (FDM) and finite element methods (FEM).

The concepts for solving partial differential equations of two methods are to replace derivative with difference quotient and curved with straight, respectively. To analyze the calculation errors of these two methods, we need to start with the approximate means of two methods.

FDM: Difference quotient instead of derivative refers to the use of Taylor's formula to expand the partial derivative into polynomial form, choosing the first few terms of Taylor's formula (or their values of operations) as the values of the partial derivative, so that the continuous partial derivative can be computed by discrete Δx and Δt, while the latter terms of Taylor's unfolding are taking, as truncation error in Δx and Δt tending toward infinitely small, truncation errors can be negligible, and the model's analog solution can be infinitely close to precise solutions, this nature is known as the convergence of difference scheme.

At the same time, there are two other errors in the difference scheme, rounding error and measurement error. The former means that the computer can only keep a limited number of decimal digits and must be rounded, thus generating rounding error. The latter is because the difference scheme is based on the initial conditions and boundary conditions, the previous Δt is used to calculate the next Δt, when errors occur in determining the initial and boundary conditions, and each measured value will be affected, resulting in measurement errors. To enable the difference scheme of the finite difference method can be used in the model calculation, except for truncation error only can be ignored when the Δx and Δt are tending to infinitely small, rounding errors and measurement errors cannot be amplified in the calculation process and controlled within a reasonable range, so the model can be adjusted by correcting boundary conditions and initial conditions. This property is called the stability of difference schemes.

FEM: In the finite element method, replacing curved with straight means that it uses element subdivision and interpolation to establish linear multiple groups of basis functions and obtain the water head value on the element by weighting. Based on the basic function of vertices, it establishes the element basis function and finally solves the first-order ordinary differential equations. The process of solving is similar to that of the finite difference method, and the first derivative is also required to be finite differentiated, and the subsequent treatment of time step is also very close to that of the finite difference method. Therefore, the error type is consistent with that of the finite difference method.

The physical meaning of FDM is relatively easy to understand and so is its error source. By optimizing Δx and Δt, the error of the finite difference method and the finite element method can be controlled, especially in the implicit difference scheme, which has absolute convergence and stability, making it flexible to adjust the time step Δt to fully consider the duration of convergence and the accuracy of the calculation.

The truncation error of differential equations for groundwater flow could be controlled, and the development of computers has allowed rounding errors to be controlled reliably, through hydrogeological conditions resurveys and correction on structural, initial and boundary conditions for conceptual and numerical models, measurement errors can be controlled. To sum up, the error of mathematical model calculation exists, but it can be effectively controlled at present to make the measurement accuracy meet the requirements.

For the advection–diffusion differential equation of material transfer in groundwater, some studies have pointed out that the difference scheme may need some correction, otherwise the relative error may exceed 75%. In the general form of advection dispersion equation (ADE) with reaction terms, besides the Taylor-like truncation series brought by the finite difference scheme, other physical terms also produce truncation errors (Notodarmojo *et al.*, 1991; Moldrup *et al.*, 1992, 1994, Atai-Ashtiani *et al.*, 1996, 1999). The truncation error has different effects on the accuracy of implicit FDM and explicit FDM numerical models in groundwater transport equations with reaction terms. The explicit finite difference method has a large deviation from the analytical solution without truncation error correction. Relative error analysis shows that after truncation error correction, the error decreases from 75% to 30%. Therefore, compared with implicit FDM, numerical truncation error correction has a significant impact on the accuracy of the groundwater migration model based on explicit FDM.

Previous studies have proposed a numerical model for the migration of chemical substances (phosphorus) in soil and groundwater with two continuous reactions (Notodarmojo *et al.*, 1991). Although the influence of zero-order and first-order truncation errors is ignored, the influence of numerical dispersion in the explicit finite difference method (FDM) is still discussed. After that, these truncation errors are quantified in the ADE with reaction item.

An application example is given to focus on the numerical truncation error formula of implicit FDM and its application on the accuracy of the numerical solution of the groundwater migration model (Figure 3.4). The numerical problem consists of a semi-infinite column, where $U = 10\,\text{cm/h}$; $D = 100\,\text{cm}^2/\text{h}$; $K = 0.5\,\text{h}^{-1}$; feed concentration $= 1000.0\,\text{mg/L}$; initial concentration $= 0.0\,\text{mg/L}$. The space increment of 20 cm and the time increment of 1.0 h are adopted.

In this study, the numerical solution is compared with the analytical solution within 24 hours. Figure 3.5 shows the comparison results. The numerical calculation results are in good agreement with the analytical solution, and the numerical truncation correction is not

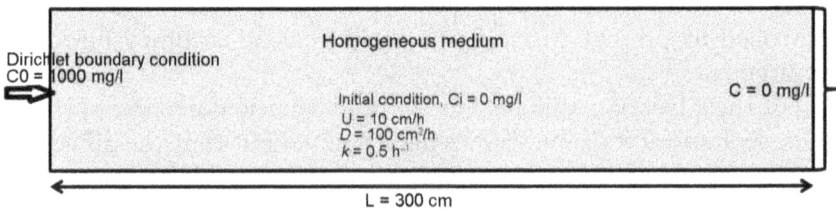

Figure 3.4. Numerical experimental devices and initial boundary conditions (Sruthi and Hyun-Su, 2016).

considered. With the increase of depth, the corrected numerical solution is closer to the analytical solution than the uncorrected. However, there is no obvious deviation between the numerical solution without implicit scheme correction and the analytical solution. Compared with the implicit FDM, the deviation between the numerical solution without numerical error correction and the analytical solution is larger for explicit FDM (Figure 3.5). In the explicit scheme, the numerical solution with error correction is in good agreement with the analytical solution.

This shows that the explicit FDM has been improved through the application of the numerical error term, while the implicit FDM has little improvement in the accuracy of the numerical solution.

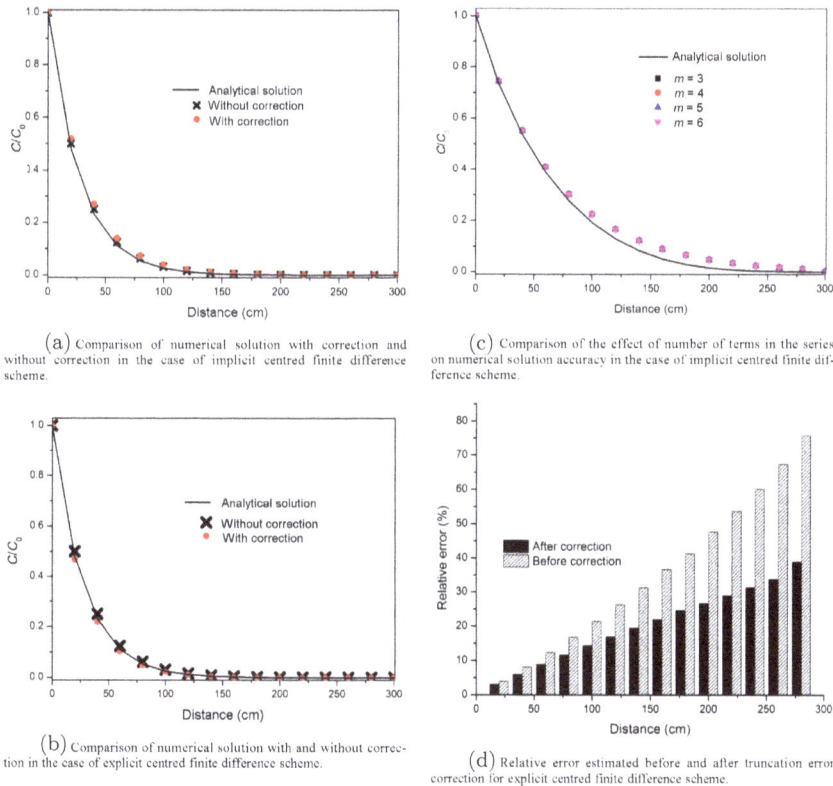

(a) Comparison of numerical solution with correction and without correction in the case of implicit centred finite difference scheme.

(c) Comparison of the effect of number of terms in the series on numerical solution accuracy in the case of implicit centred finite difference scheme.

(b) Comparison of numerical solution with and without correction in the case of explicit centred finite difference scheme.

(d) Relative error estimated before and after truncation error correction for explicit centred finite difference scheme.

Figure 3.5.   Comparison of model simulation results and analytical solutions before and after correction and calculation of relative errors (Sruthi and Hyun-Su, 2016).

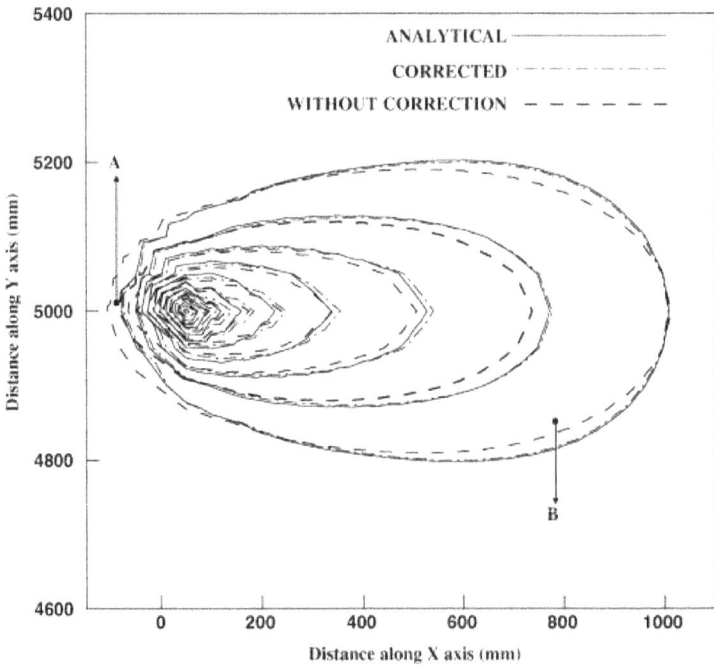

Figure 3.6.   Comparison of the solution of the convection–diffusion numerical model with adsorption term before and after correction with the analytical solution (Ataie-Ashtiani and Hosseini, 2005).

Also, the study illustrates the influence of increasing the number of terms used in series (m) on the accuracy of the solution in the case of implicit FDM. The number of terms used in the series (m) controls the error between the numerical solution and the analytical solution. The truncation error can be minimized by increasing the number of terms in the series. However, at a certain m value, the numerical solution converges completely with the analytical solution (Figure 3.6). Therefore, increasing its value after that value does not significantly improve the accuracy of numerical results.

The relative errors of different FDMs truncation error corrected and uncorrected numerical solutions are estimated. The research results show that by removing the truncation error in FDM, the numerical error will be greatly reduced (Figure 3.6). After truncation error correction, the maximum error is reduced from 75% to 30% (Figure 3.6). Therefore, it is of great significance to study the truncation error correction of FDM, because FDM is used to solve numerical problems in groundwater models such as MT3D, and the application of truncation error correction term can reduce the error of these FDM numerical results (Sruthi and Hyun-Su, 2016).

Not only will the reaction term bring additional errors to the numerical solution of the convection–diffusion differential equation of groundwater system but also the adsorption effect will affect the numerical error of its finite difference solution (Ataie-Ashtiani and Hosseini, 2005).

In this study, the numerical error of the finite difference solution of the two-dimensional convection–diffusion equation with linear adsorption is obtained from Taylor analysis, and it is eliminated from the numerical solution. The results show that the Crank–Nicolson method based on truncation error analysis is the most accurate. By comparing with the analytical solution for calculating the distribution of pollutant plume in the uniform flow field, the influence of these truncation errors on the numerical solution of two-dimensional first-order reaction convection–diffusion equation is illustrated (Figure 3.6). Eliminating these errors can improve the numerical results and reduce the difference between the numerical solution and the analytical solution (Ataie-Ashtiani and Hosseini, 2005).

In addition to modifying the finite difference scheme of the convection–diffusion equation for groundwater flow, some studies also use the finite volume element method to obtain its numerical solution and give the corresponding error estimation. Compared with the finite difference method, the calculation accuracy is improved (Sinha and Geiser, 2007; Zhang, 2009).

Chapter 4

The Complexity of Groundwater Systems: Scientific Research Methods

4.1 Characterization of Functional Complexity of the Groundwater System

4.1.1 *Groundwater order parameter*

The groundwater system is an organism of resource subsystem, ecological subsystem, and geological environment subsystem and is an orderly dissipative structure. In the dissipative structure, the order parameters, which dominate the changes of other variables and further control the overall evolution process of the system, determine the order degree of the system, so the order degree and evolution direction of the system can be expressed by the order parameters. The orderliness of the groundwater system function is that the functions of resources, ecology, and geological environment subsystems in the system can maintain a certain order so that the geological environment subsystem is stable and the ecological subsystem is balanced, and finally the virtuous circle and healthy evolution of groundwater system can be realized. The function of groundwater resources refers to the supply guarantee function or effect of groundwater resources with certain recharge, storage, and renewal conditions in the groundwater system. The function of resources can be expressed from the aspects of groundwater recharge, available amount, the change rate of groundwater level, the balance of exploitation and recharge, hydraulic conductivity of aquifer, exploitation degree, etc.

The groundwater ecological function refers to the support of groundwater on land surface vegetation, lakes, wetlands, or land quality. The function can be expressed from the degree of groundwater level deviating from ecological water level, groundwater mineralization degree, soil salinity, soil moisture content, vegetation coverage rate, land desertification rate, soil salinization rate, etc. The geological environmental function of groundwater refers to the supporting and protecting effect of the groundwater system on the stability of the geological environment. The geological environmental function can be expressed from the aspects of land subsidence rate, degree of groundwater level lower than the stable control water level, seawater intrusion, rate of groundwater depression cone area, etc. The selection of order parameters for groundwater function evaluation should be based on the main environmental geological problems in the study area and the availability of data.

The following examples are based on the groundwater function in Puyang City, to show the main influencing factors and major environmental geological problems characterizing groundwater function, the order parameters their meanings, and calculation methods (Table 4.1).

According to the actual situation in northern China and referring to relevant standards and previous research results (Zhang *et al.*, 2003; Sun and Liu, 2009; Cheng *et al.*, 2018), each order parameter is divided into five levels, and the classification standard is listed in Table 4.2.

4.1.2　*System order degree*

The groundwater system is a dissipative structure, and the phase change of the system could lead the system toward new orderly or disorderly. So order degree of groundwater system functions is introduced.

Take into account the groundwater subsystem S_k. Let the order parameter in the process of its development be $x_k = (x_{k1}, x_{k2}, \ldots, x_{kn})$, where $n \geq 1$. To calculate the order degree of groundwater subsystem S_k and order parameter x_{ki}, this article uses the idea of fuzzy mathematics, computes order parameters by fuzzy affiliation, that is, firstly computes the membership degree of the order parameter x_{ki} relative to the various levels of groundwater function,

Table 4.1. Functional order parameters of groundwater.

Functional category	Order parameters		Meaning	Calculation formula
	Name	Comments		
Resource function (B_1)	Modulus of recharge	Groundwater recharge per unit area	Indicates the recharge capacity of groundwater resources	Modulus of recharge $= \dfrac{\text{Recharge of groundwater}}{\text{Evaluation area}}$
	Available resource modulus	Available groundwater per unit area	Indicates the amount of available groundwater resources	Modulus of available resources $= \dfrac{\text{Available groundwater}}{\text{Evaluation area}}$
	Groundwater level change rate	Variation of groundwater level per unit time	Indicates the renewal capacity of groundwater resources	Rate of groundwater level change $= \dfrac{\text{Change in groundwater level}}{\text{Evaluation of time}}$
	Balance rate of production and compensation	The ratio of production to recharge	Indicates the reproducibility of groundwater resources	Balance rate of production and compensation $= \dfrac{\text{Amount of groundwater mining}}{\text{Recharge of groundwater}}$
	Degree of groundwater exploitation and utilization	The ratio of groundwater exploitation to availability	Indicates the remaining exploitation potential of groundwater resources	Degree of groundwater exploitation and utilization $= \dfrac{\text{Amount of groundwater mining}}{\text{Available groundwater}}$

(*Continued*)

Table 4.1.   (*Continued*)

Functional category	Sequence parameters			
	Name	Comments	Meaning	Calculation formula
Ecological functions (B_2)	Groundwater mineralization	Salt content per unit volume of groundwater	Reflecting the effect of groundwater on ecological environment from groundwater mineralization	Degree of mineralization of groundwater $= \dfrac{\text{Salt content of groundwater}}{\text{Groundwater volume}}$
	Land desertification rate	The ratio of desertification area to the total area	Reflecting the effect of groundwater on ecological environment from the degree of land desertification	Desertification ratio $= \dfrac{\text{Land desertification area}}{\text{Total area}}$
	Soil salinization rate	The ratio of soil sanitization area to the total area	Reflecting the effect of groundwater on ecological environment from the scope of soil salinization	Soil salinization rate $= \dfrac{\text{Soil salinized area}}{\text{Total area}}$

Geological environment function (B₃)	Land subsidence rate	Land subsidence per unit time	Reflecting the effect of groundwater on geological environment stability from land subsidence status	$\text{Land subsidence rate} = \dfrac{\text{Land subsidence area}}{\text{time}}$
	Groundwater depression rate	The ratio of groundwater depression area to total area	Reflecting the effect of groundwater on the stability of the geological environment	$\text{Groundwater depression rate} = \dfrac{\text{Groundwater depression area}}{\text{Total area}}$

Table 4.2.   Grading standard of groundwater functional order parameters.

Functional Category	Order Parameters	Grading Criteria				
		Level 1 Excellent	Level 2 Good	Level 3 Moderate	Level 4 Poor	Level 5 Extremely poor
Resource Function (B1)	Modulus of recharge/10000 $m^3 \cdot km^{-2}$	>35	25~35	15~25	5~15	<5
	Available resource modulus/10000 $m^3 \cdot km^{-2}$	>25	15~25	8~15	4~8	<4
	Groundwater level change rate/$m \cdot a^{-1}$	<0.01	0.01~0.5	0.5~1.0	1.0~1.5	>1.5
	Balance rate of production and compensation/%	<10	10~30	30~60	60~90	>90
	Degree of groundwater exploitation and utilization/%	<30	30~80	80~100	100~120	>120
Ecological function (B2)	Groundwater mineralization/$g \cdot L^{-1}$	<1	1~3	3~5	5~10	>10
	Land desertification rate/%	<2	2~4	4~6	6~8	>8
	Soil salinization rate/%	<2	2~4	4~6	6~8	>8
Geological Environment Function (B3)	Land subsidence rate/$mm \cdot a^{-1}$	<1	1~2	2~5	5~10	>10
	Groundwater depression rate/%	<5	5~10	10~15	15~20	>20

and then taking the score of each functional level as a weight, to calculate the order degree of parameters, where the affiliated function takes the halved trapezoidal distribution function.

Let the k-th groundwater subsystem S_k has m order parameters $(k = 1, 2, 3)$, and the i-th order parameter is x_{ki}. The standard value of grade j of groundwater function of this order parameter is C_{ij} $(i = 1, 2, \ldots, m; j = 1, 2, \ldots, n)$.

If x_{ki} is the reverse indicator, the affiliation function for each function rating of x_{ki} is:

$$\gamma = \begin{cases} 1 & x_{ki} \leq c_{i1} \\ \dfrac{c_{i2} - x_{ki}}{c_{i2} - c_{i1}} & c_{i1} < x_{ki} \leq c_{i2} \\ 0 & x_{ki} > c_{i2} \end{cases} \tag{4.1}$$

$$\gamma_{ij} = \begin{cases} 1 - \dfrac{c_{ij} - x_{ki}}{c_{ij} - c_{ij-1}} & c_{ij-1} \leq x_{ki} \leq c_{ij} \\ \dfrac{c_{ij+1} - x_{ki}}{c_{ij+1} - c_{ij}} & c_{ij} < x_{ki} \leq c_{ij+1} \\ 0 & x_{ki} > c_{ij+1} \ or \ x_{ki} < c_{ij-1} \end{cases} \quad j = 2, \ldots, n-1 \tag{4.2}$$

$$\gamma_{in} = \begin{cases} 1 & x_{ki} \leq c_{in-1} \\ \dfrac{c_{in} - x_{ki}}{c_{in} - c_{in-1}} & c_{in-1} < x_{ki} \leq c_{in} \\ 0 & x_{ki} > c_{in} \end{cases} \tag{4.3}$$

If x_{ki} is the forward indicator, the affiliation function for x_{ki} of each functional evaluation level is:

$$\gamma_{i1} = \begin{cases} 1 & x_{ki} \geq c_{i1} \\ \dfrac{c_{ki} - x_{i2}}{c_{i2} - c_{i2}} & c_{i2} \leq x_{ki} < c_{i1} \\ 0 & x_{ki} < c_{i2} \end{cases} \tag{4.4}$$

$$\gamma_{ij} = \begin{cases} 1 - \dfrac{x_{ki} - c_{ij}}{c_{ij-1} - c_{ij}} & c_{ij} \leq x_{ki} \leq c_{ij-1} \\ \dfrac{x_{ki} - c_{ij+1}}{c_{ij} - c_{ij+1}} & c_{ij+1} \leq x_{ki} < c_{ij} \\ 0 & x_{ki} > c_{ij-1} \ or \ x_{ki} < c_{ij+1} \end{cases} \quad j = 2, \ldots, n-1 \tag{4.5}$$

$$
\gamma_{in} = \begin{cases} 0 & x_{ki} \geq c_{in-1} \\ 1 - \dfrac{x_{ki} - c_{in}}{c_{in-1} - c_{in}} & c_{in} \leq x_{ki} < c_{in-1} \\ 1 & x_{ki} < c_{in} \end{cases} \tag{4.6}
$$

The order degree $u_{ki}(x_{ki})$ for order parameter x_{ki} is:

$$
u_{ki}(x_{ki}) = \gamma\omega = \sum_{j=1}^{n} \gamma_{ij}\omega_j \quad i = 1, \ldots, m \tag{4.7}
$$

In the formula, $u_{ki}(x_{ki})$ is the order degree of the order parameter x_{ki}; ω_{ki} is the score of each groundwater function grade for x_{ki}, due to the order parameters is divided into five grades, so $\omega = [0.999\ 0.75\ 0.5\ 0.25\ 0.001]$.

$u_{ki}(x_{ki}) \in (0,1)$ and the greater the value, the greater contribution to the order degree of subsystem S_k from x_{ki}. The order degree from all order parameters $x_{ki}(i = 1, 2, \ldots, m)$ to subsystem S_k is

$$
u_k(x_k) = \sum_{i=1}^{m} \lambda_i u_{ki}(x_{ki}) \quad \lambda_i \geq \sum_{i=1}^{m} \lambda_i = 1 \tag{4.8}
$$

In this formula, λ_i the weight coefficient of the order parameter x_{ki}, represents the role of the parameter x_{ki} in keeping the subsystem running in order status.

4.1.3   *Groundwater function and its grade*

Due to $u_k(x_k) \in (0,1)$ and the larger $u_k(x_k)$, the greater "contribution" to the order degree of the subsystem from x_k, the higher the order degree of the subsystem, and the stronger the groundwater function. Therefore, the order degree of the subsystem F_k can be used to express the strength of the groundwater function of the subsystem, and the expression is:

$$
F_k = u_k(x_k) \tag{4.9}
$$

In the formula, F_k is the groundwater function, $k = 1, 2, 3$, respectively, to represent the subsystems of resource function, ecological function, and geological environment function.

Due to the limited function of the groundwater system in a specific period, the order parameters among subsystems compete with each other, and the order degree in subsystems cannot increase at the same time. The improvement of the order degree in one subsystem may lead to a decrease in the other subsystems, making it impossible to determine how the order degree of the system changes. The entropy theory of dissipative structure provides the basis for solving this problem, although system evolution cannot be quantitatively calculated based on entropy and it is not easily expressed by explicit functions, the relationship between entropy and order degree can be used to qualitatively analyze the direction of system evolution, i.e., entropy decreases, ordering enhances, groundwater system function enhances, entropy increases, ordering weakens, and groundwater function decreases.

Based on the definition of information entropy, ordered information entropy F representing the functional size of a groundwater system can be established using the parameter order degree, whose function is

$$F = -\sum_{k=1}^{3} \frac{1 - F_k}{3} \log \frac{1 - F_k}{3} \tag{4.10}$$

where F is the total information entropy of the groundwater system.

According to the functional order degree and information entropy results of the groundwater system, and taking into account $0 < F_k < 1$ and $0 < F < 0.5$, the groundwater functional classification standards are listed in Table 4.3.

Table 4.3. Grading standards for groundwater target function evaluation.

Functional level	I	II	III	IV	V
Functional status	Excellent	Good	Moderate	Poor	Extremely poor
F Resource function (B1)	≥0.8	0.6–0.8	0.4–0.6	0.2–0.4	<0.2
Ecological function (B2)	≥0.8	0.6–0.8	0.4–0.6	0.2–0.4	<0.2
Geological environment (B3)	≥0.8	0.6–0.8	0.4–0.6	0.2–0.4	<0.2
Comprehensive function (A)	≤0.24	0.23–0.35	0.35–0.42	0.42–0.46	>0.46

4.2   Theory of Dissipative Structure

Prigogine's dissipative structure theory reveals the mechanism of system evolution under certain external conditions, which has an important methodological significance (Wu, 1995). The contribution of conclusions about system non-equilibrium, open and ordered, non-linear action, etc., based on dissipative structure theory is remarkable to the theory of nonlinear system (Guan, 1999), which lays a rigorous scientific basis for the further development of systems science (Chi, 2002). Since creation, the theory has been widely used in the fields of biological ecology, geological sciences, geophysics, and social sciences (Chen, 2001).

The evolution of groundwater systems is the process by which groundwater systems replace old macrostates with new macrostates under the influence of natural or human factors. The evolution of the groundwater system consists of a causal chain, which shows the transformation from a stable state to an unstable state, and then to a new stable state (Liao, 1997). All spontaneous processes in nature are irreversible, so all the states experienced in the evolution process are non-equilibrium states, and non-equilibrium structures can be formed in the evolution process of non-equilibrium states (Shen, 2001). The water resource system is a dissipation structure (Chang *et al.*, 2002), and the process of water cycle evolution is a dissipation process (Zhang *et al.*, 2001). The evolution of a groundwater system is the evolution of a nonlinear complex system. The groundwater system has a wide exchange of matter, energy, and information with the outside world, and it is a complex open system that is neither isolated nor closed. According to Prigogine, when a sustained exchange of energy and mass occurs between the open system and the environment, it will be possible for the system to be moved away from the equilibrium state, and that the consumption of the system's energy due to irreversible processes can make it "self-organized" and generate temporally and spatially ordered "dissipation structure". Based on dissipative structure, the evolution mechanism of groundwater system under the influence of human activities can be discussed to provide a new research idea for groundwater system analysis.

4.2.1 Theory of dissipative structure

Dissipative structure refers to an open system far from equilibrium. When its variation reaches a certain threshold, abrupt changes may occur through fluctuations, from the original chaotic state to a state of spatial, temporal, and functional order, and this new stable ordered structure is formed in the nonlinear region far from equilibrium, so it is necessary to exchange matter and energy with the external environment to maintain it. This stable and orderly structure is called a dissipative structure, that is, a dynamic and orderly system generated under non-equilibrium conditions that rely on the continuous input and output of matter, energy, and information to maintain its internal nonlinear interaction.

The dissipation structure theory holds that in a system with irreversible processes, its entropy change (ds) consists of two parts, $d_i s$, and $d_e s$. $d_i s$ is the entropy generation of irreversible processes in the system, which always increases the entropy of the system, i.e., $d_i s \geq 0$. $d_e s$ is the entropy flow provided by the outside, which can be positive or negative. The entropy change ds of the system satisfies $ds = d_i s + d_e s$. When $ds < o$, the system tends to a certain spatio-temporal order state. At this time, under specific conditions, the system will form a dissipative structure. The four basic characteristics of dissipative structure system are as follows: (1) dissipative structure system is an open system; (2) the elements and subsystems in the dissipative structure system are in a non-equilibrium state, and the system is a non-equilibrium "living" orderly structure; (3) the interaction of various elements and subsystems in the dissipative structure system presents a nonlinear mechanism; (4) there is a competition mechanism among the subsystems in the dissipative structure system, and the whole system is characterized by fluctuation. The opening of the dissipative structure system makes it possible to exchange material, energy, and information with the environment. The competition and coordination of each subsystem in the system are the embodiment of the non-equilibrium and nonlinearity of the system, and the essence of competition and coordination is the competition and coordination of input materials.

4.2.2   Dissipative structure in the evolution of groundwater systems

The groundwater system is a complex of several independent units, which are controlled by various natural and human factors, have different levels of interrelation and interplay, and have four-dimensional properties and respective characteristics in space and time, and continuously evolve. The evolution of groundwater system is largely controlled by surface water input and output systems (Chen and Ma, 2002). The groundwater system has multi-attribute functions of resources, ecology, environment and regulation, and storage. The motivation of the interaction and restriction between these multi-attribute functions is the renewal capacity of regional groundwater and the evolution of the water circulation (Zhang *et al.*, 2004).

With the advent of nonlinear system dynamics and dissipative structure theory, it is possible to examine the evolution of groundwater system with a brand-new thinking. A key problem is to understand the evolution of groundwater system is a nonlinear complex system evolution process. The following discusses this problem from several aspects:

1. *Groundwater system is a complex open system*: The groundwater system is closely related to the environment. Even in the deep crust, the physical and chemical interaction between water and surrounding rock, and the adaptation of medium structure to in-situ stress all show that groundwater can't exist without environment, and it will transmit information in the long-term evolution of geological history and even under the action of the solid tide. First of all, the environment inputs materials, energy, and information to the groundwater system and at the same time receives various outputs from the groundwater system, thus the groundwater system has the movement of material and information, which can maintain its existence. Secondly, the environment affects the structure, function, and order of the groundwater system in the process of formation, development, and extinction. The groundwater system is therefore a complex open system where there is a continuous and extensive exchange of matter, energy, and information with the environment.

2. The evolution of the groundwater system is an irreversible process in a non-equilibrium and ordered state: The sign that the system reaches the equilibrium state is that all elements are uniform and single, disordered, with extremely large entropy value, and the greatest degree of chaos. The equilibrium system is in a chaotic state with an extremely large entropy value, and it is impossible to produce a new ordered structure. The imbalance is a source of order, and instability is the driving force to produce new structures (Yang *et al.*, 1998).

The groundwater system is an entity that has been gradually formed and continuously modified during the development of a long geological history, and it has never reached an equilibrium that is no longer changing over time. The elements that make up the groundwater system are the products of the natural dynamic process. For a long period, the groundwater system has different manifestations at different times, and it also shows spatial changes. Therefore, considering geological time as a time scale, the groundwater system is a non-equilibrium structure that is ordered in space, time and function. With the continuous progress of hydrogeological construction and reconstruction, the groundwater system undergoes the evolution of the water circulation system, seepage field, and hydrochemical field. Irreversible factors such as karstification, tectonic evolution (especially neotectonic movement), climate, and environmental changes determine that the evolution of the groundwater system is irreversible. Thus, the evolution of groundwater system is an irreversible process in which it continues to move away from equilibrium toward order.

3. The interaction between the subsystems of the groundwater system is nonlinear: Nonlinear systems is inhomogenous, asymmetric, and coherent (Qin, 1994). The various subsystems that make up the groundwater system are interconnected and interacting. Changes in the subsystem will eventually affect the groundwater system. The overall characteristics and activities of the groundwater system rely on each subsystem and are manifested in their mutual connection, function, and restriction. However, the sum of their independent characteristics and activities cannot reflect the overall characteristics and activities of the system. The synthesis of these subsystem actions cannot be described by linear equations and it is usually a nonlinear coupling action. The nonlinear interaction among the subsystems in the system, is one of the necessary conditions for the continuous evolution of the groundwater system toward orderliness.

4. Human activities have a significant influence on the evolution of the groundwater system: The temporal and spatial distribution and evolution of groundwater systems are not only restricted by natural factors but also affected by the social environment, especially human activities. The input of the material, energy, and information from human activities to the groundwater system controls the evolution of the groundwater system.

The fluctuation is a major hallmark of the evolution of groundwater systems, and it is intrinsically different from traditional groundwater dynamics (Xu *et al.*, 2002). The groundwater system has stable state and unstable state. The premise of a stable state is that the environmental conditions do not change with time. When the input and output changes slightly, the groundwater system can be considered to be in a stable state. Once the groundwater system is disturbed by human activities, it will become a natural–artificial composite system. As a new environmental factor, activities such as exploitation of groundwater will increase the fluctuation of the groundwater system, resulting in instability of the groundwater system. The state variables of the groundwater system show normal fluctuation under the macroscopic stable state condition and abnormal fluctuation under the unstable state condition.

The intensity of human activities determines the nature and speed of groundwater system evolution. For example, artificial exploitation and recharge of groundwater, surface and underground engineering activities, and pollution have gradually replaced or weakened the natural source and sink items of groundwater. The flow field, the spatial structure, and the hydrochemical field have successively entered abnormal fluctuations, and the groundwater system will evolve to a new state. For open groundwater systems far from equilibrium, it is a nonlinear and non-equilibrium structure, moreover, the nonlinear action among subsystems may amplify the small fluctuation to form a huge fluctuation, and make the subsystems with nonlinear action spontaneously produce synergistic action, thus generating orderly and stable self-organization dissipative structure, releasing energy suddenly, generating deformation and environmental geological disasters. Water environmental problems such as regional groundwater depression cone, land subsidence, and salinization of deep groundwater are the products of abnormal fluctuation of groundwater system under the influence of human activities, which evolves from previous stable state to unstable state.

4.3   Characterization of Uncertainty in Groundwater System

4.3.1   *Uncertainty of groundwater systems*

Information features are the unified embodiment of the elements, structure, and function of the system. Without information features, the system cannot be described (Zuo and Ma, 1994). Taking the perspective of uncertainties, uncertain information of the groundwater system can be divided into four categories: randomness, fuzziness, greyness, and unknown, which can be described as follows:

1. *Randomness*: Randomness is that there is no inevitable causal relationship between objective conditions and the occurrence of events, thus the occurrence of events also presents uncertainty, which is mainly caused by factors such as insufficient conditions for the occurrence of events or interference of accidental factors. For example, the variation of water-bearing property, water abundance, and hydrogeological parameters of groundwater system has obvious spatial and time variability.

2. *Fuzziness*: Fuzziness originates from the fuzziness of the concept of the thing itself. Due to the incompleteness of knowledge and the complexity of the system, there is no clear "boundary" for the definition of things, that is, it is difficult to clarify whether a thing belongs to a certain concept. For example, the concepts of permeable layer, weak permeable layer, or impermeable layer are relative, and the same problems also exist in the confining boundary, recharge boundary, and constant head boundary.

3. *Greyness*: Due to the complexity of things and the limitation of other factors, it is impossible to fully grasp the information contained in the system. People only know a part of the information. To a certain extent, this uncertainty, which is partially known and partially unknown, is called greyness. For example, in the process of modeling, some problems such as the unclear concept of parameters or unclear physical meaning may be encountered, and the model built on this basis will show greyness.

4. *Unknown*: Unknown refers to the uncertainty that makes it difficult to objectively understand and describe according to the information mastered and can only be started from the perspective of pure subjectivity. The groundwater system is a complex and open giant system. It is difficult to make a complete understanding on it, sometimes it is not necessary to make a complete understanding, and sometimes it can only be judged subjectively (Lei, 2018).

When analyze the complex groundwater systems, we should combine the random uncertainty method, fuzzy uncertainty method, grey theory, and unascertained theory effectively, to accurately and effectively characterize the groundwater system according to the research content.

4.3.2   *Random uncertainty method*

The random uncertainty method is mainly based on probability theory and mathematical statistics, to consider the random uncertainty of the groundwater system. The application range is wide and the theory is mature. According to the solution principle, the random uncertainty analysis methods for groundwater systems can be divided into three categories: the Monte Carlo method, first-order second-moment method, design point method. These can be described as follows:

1. *Monte Carlo method*: The Monte Carlo method was developed by the United States in the 1940s for the "Manhattan Plan" of atomic bombs in World War II. The name Monte Carlo is used to express the nature of the random (gambling) phenomenon. In 1965, Warren and Price first applied the Monte Carlo model to groundwater research. In the uncertainty analysis of the groundwater system, the Monte Carlo method is a probability method with the highest accuracy.

The Monte Carlo simulation method assumes that the probability distribution function of the random variable is known, many groups of input variables are obtained by random sampling method, each group of variables is equivalent to a statistical test, and then the random variables are brought into the model to obtain a large number of outputs. According to the statistics of the output results, statistical estimators such as mean value and variance can be obtained, and the probability distribution of the output results can be fitted.

The main ideas of the Monte Carlo simulation method are as follows:

(1) Random sampling is carried out in the feasible region of each random variable or according to their probability distribution. In the sampling process, the Latin hypercube sampling method can be adopted to make the sample coverage rate higher and reduce the sampling times.

(2) After randomly combining the sampling results of random variables, the corresponding output results are obtained through the model.

(3) Statistically analyze the model output results, according to estimators such as the mean and variance to fit the probability distribution of output results, which quantitated describes the uncertainty of the model (Zhang *et al.*, 2018).

The Monte Carlo method avoids the mathematical difficulties in random analysis, no matter whether the groundwater model is linear or not, whether the random variables are normal or not, as long as the number of simulations is enough, a accurate probability distribution can be obtained, and the convergence speed is independent of the dimension of the problem, and the program structure is simple. However, some shortcomings of the Monte Carlo method cannot be ignored, such as slow convergence speed and difficult to estimate and control calculation errors.

2. *Generalized Likelihood Uncertainty Estimation (GLUE)*: In 1992, Beven and Binley first explicitly proposed the phenomenon of equifinality for different parameters in the uncertainty analysis of hydrological models and generalized likelihood uncertainty estimation (GLUE) based on the regionalized sensitivity analysis. An important idea of the GLUE method is to abandon the concept of global optimal parameter solution and think that it is not a single model parameter, but the combination of model parameters that leads to the performance of the model. Firstly, in the preset parameter distribution space, the combination of parameter values of the model is randomly selected according to the prior distribution, and the model is run by the Monte Carlo method. Secondly, the appropriate likelihood function is selected to evaluate the proximity between the operation results of the model and the actual observation. The closer to the actual observation results, the higher credibility of the considered parameter group, and a likelihood weight is given by the likelihood function. When the likelihood value of the simulation is lower than a specified likelihood critical value, the parameter group is considered unacceptable, and its likelihood value is set to 0. Finally, according to the reserved parameter group and the corresponding simulation results, all likelihood values are normalized to obtain the posterior probability distribution of simulation and model parameters.

The advantages of the GLUE method are simple principle, convenient operation, and strong reliability, which can be applied to various uncertainty analyses. GLUE considers all sources of errors in the simulation process through the likelihood function and can use Bayesian statistical derivation to obtain model parameters and the posterior distribution.

3. *Markov Chain Monte Carlo Method (MCMC)*: Markov chain is a discrete-time stochastic process with Markov properties mathematically and has the properties of the irreducible, aperiodic, and stationary distribution. In the Bayesian statistical framework, the MCMC method establishes a Markov chain $\pi(x)$ with stationary distribution and random sample in this smooth stationary distribution, then sufficient search of the probability distribution space of the target function during the evolution of Markov chains. The MCMC method can continuously adjust the search strategy by combining the previous sampling information, fully sample the high probability density region, and converge to the posterior distribution of the objective function (Blasone *et al.*, 2008; Hassan *et al.*, 2009; Vrugt *et al.*, 2003).

The sampling algorithm is the core of the MCMC method, which determines the sampling efficiency and stability of the MCMC method. Different types of MCMC methods can be established to construct sampling algorithms for specific research objectives (Vrugt *et al.*, 2008), such as Metropolis–Hastings algorithm, single component Metropolis–Hastings algorithm, Gibbs sampling algorithm, adaptive proposed distribution sampling algorithm, and so on (Pan, 2016). The following provides a brief introduction to commonly used sampling algorithms:

(i) *Metropolis–Hastings algorithm*: The Metropolis–Hastings (MH) algorithm specifies simple transfer nuclei to generate Markov chain of $\pi\theta$ with invariant distributions, that can be considered as sample of $\pi\theta$. The MH algorithm is based on the Accept—Denial method: generates new candidate values θ based on the probability of acceptance α from the suggested distribution $q(\bullet|\theta)$. The MH algorithm is divided into the following steps:

    (1) Initialization: Select an initial point θ_0; set $\theta_{old} = \theta_0$; set $chain(1) = \theta_0$, and $i = 2$;

    (2) Select a new candidate value from the suggested distribution: $\theta^* \sim q(\bullet|\theta_{old})$;

    (3) Receive new candidate values with a probability of acceptance α, $\alpha = \min(1, \frac{\pi(\theta^*)q(\theta_{old}|\theta^*)}{\pi(\theta_{old})q(\theta^*|\theta_{old})})$; set if accepted $chain(1) = \theta^*$, and $\theta_{old} = \theta^*$, if rejected $chain(i) = \theta_{old}$;

    (4) Set $i = i + 1$ and perform step (2) again, cycle the steps in turn.

It is assumed that suggested distribution is symmetrical, that is if $q(\theta^*|\theta_{old}) = q(\theta_{old}|\theta^*)$, the MH algorithm is the Metropolis algorithm proposed by Metropolis. In Bayesian frameworks, the posterior distribution $\pi(\theta)$ contains normalization constants, but in the MH algorithm, the normalization constant is counterbalanced.

(ii) *SCMH algorithm*: In the MH algorithm, all elements in the Markov chain are updated at the same time. When a Gaosrecommended distribution is adopted, it means that sampling from a multivariate Gauss distribution is required. In the single component (SC) MH algorithm, the Markov chain is updated element by element, θ^i is the t-iteration sampling value for the element, which means multiple suggested distribution and probability of acceptance under different conditions. The suggested distribution of θ_t^i can be a univariate normal distribution that is centered on the previous sampling point θ_{t-1}^i, with some fixed σ_i^2 variance, that is, $q_t^i \sim N(\theta_{t-1}^i, \sigma_i^2)$. A posteriori probability density function is required for probability of acceptance calculations, and the difference from the MH algorithm is that when sample θ^i, the previous sampled $\theta^1, \theta^2, \ldots, \theta^{i-1}$ are needed to calculate the posterior probability density function. In multidimensional inverse problems, the recommended distribution of the SCMH algorithm is always simple, making it easy to sample in the posterior probability density function. The difference of SCMH algorithm from the MH algorithm is mainly in steps (2) and (3).

(2) Sampling $\theta_i^* \sim q_t^i$;
(3) The probability of acceptance is

$$\alpha = \min\left(1, \frac{\pi(\theta_t^1, \theta_t^2, \ldots \theta_t^{i-1}, \theta_i^*, \theta_{t-1}^{i+1}, \ldots, \theta_{t-1}^n)}{\pi(\theta_t^1, \theta_t^2, \ldots \theta_t^{i-1}, \theta_{t-1}^i, \theta_{t-1}^{i+1}, \ldots, \theta_{t-1}^n)}\right);$$

if accepted, set $\theta_t^i = \theta_i^*$; if rejected, set $\theta_t^i = \theta_{i-1}^i$.

The SCMH algorithm has its inherent disadvantages. Because one-dimensional suggested distribution does not consider the correlation among parameters, while the covariance matrix in multivariate Gaussian distribution contains correlativity, the mixed attribute of the result is poor. One way to solve the unknown correlation is to rotate the suggested distribution on some predefined steps, which can be completed by calculating the covariance matrix of the currently generated chain and the principal components of the covariance matrix.

(iii) *Gibbs algorithm*: Gibbs sampling algorithm is a correction of the SCMH algorithm and is widely used in different situations. The Gibbs algorithm uses a one-dimensional full-conditional posterior distribution $\pi(\theta_i|\theta_1, \ldots, \theta_{i-1}, \theta_{i+1}, \ldots, \theta_p)$ for sampling one by one. Gibbs sampling assumes that these conditional distributions are known, that is, elements except those to be sampled are known. Candidate values generated in Gibbs sampling are always accepted. The Gibbs algorithm can be used when conditional distributions can be easily sought. The difference of Gibbs algorithm from the MH algorithm is mainly in steps (2) and (3);

(2) Sampling $\theta_j^i \sim \pi(\theta_i | \theta_1, \ldots, \theta_{i-1}, \theta_{i+1}, \ldots, \theta_p)$;

(3) Setting $Chain_i(j) = \theta_j^i$.

However, the acquisition of one-dimensional conditional distributions is complex. If the analytical form of the conditional distribution is unknown, it can be obtained from calculating the empirical distribution of the target distribution $\pi(\theta)$.

MCMC's sampling algorithm continues to be improved due to the development of mathematical statistical methods. The advantages of the MCMC method are simple principles, flexibility, and uncertainty analysis that can be applied to various water environment models. In particular, the MCMC method has good adaptability to complex nonlinear, high-dimensional, and multi-peak distribution parameter uncertainty problems. The disadvantage of the MCMC method is that it needs the strong computing power and time-consuming. For complex high-dimensional nonlinear problems, a large number of Monte Carlo simulation times are often required to obtain simulation convergence. Also, due to the logical computing characteristics of the MCMC method, it has great restrictions on the application of technologies such as parallel computing (Zeng, 2012).

4. *First-Order Second-Moment (FOSM)*: Dettinger and Wilson first put forward the basic idea of applying the first-order second-moment (FOSM) method to solve groundwater problems in 1981. Townley and Wilson generalized the method in 1985, incorporating the random effects in nonconstant flow and storage coefficients and boundary conditions. Wang and Hsu (2013) analyzed pore elasticity problems using the means and covariance functions of hydraulic conductivity, pore water pressure, and displacement at constant load and flux boundary based on FOSM method. The results of the FOSM pore elasticity model were compared with that of MC simulations, and the results of the two random methods showed similar laws.

The first-order second-moment method is to predict the probability distribution characteristics of output variables according to the probability distribution characteristics of input variables. If there is an obvious correlation between the input and output variables, the probability distribution of the output variables can be predicted by a simple calculation. Otherwise, a virtual experiment (or simulation) can be used to predict the probability distribution characteristics of the output variable that is a function of the input variable.

The concept of the FOSM approach can be summarized as follows (Kim *et al.*, 2020):

(1) When the probabilistic distribution feature of the input $(\mu_x; \sigma_x)$ is known, and the probability distribution characteristics of the $(\mu_x; \sigma_x)$ could be output by Taylor series;
(2) D_x is the fraction of the differential, used as a sensitivity measure for output to the input;
(3) Estimated output $C_v(C_v = \mu_x/\sigma_x)$ with the results of the FOSM method is the uncertainly measurement of output;
(4) Based on the FOSM method, the uncertainty of output variables could be compared when considering each input variable or considering all inputs at the same time.

Compared with the Monte Carlo method, the FOSM method has higher computational efficiency. It predicts the distribution of output variables according to all possible combinations of input variables. Wang and Hsu (2013) applied the first-order second-order moment method to the unconditional and conditional analysis of the equation for groundwater flow and solute transport and compared the results with the corresponding Monte Carlo simulation results. The FOSM method successfully reproduces the Monte Carlo results with 30–60 times less computing time.

FOSM method also has its limitations: If the correlation between input variables and output variables is weak, the FOSM method cannot be widely used for accurate sensitivity analysis due to its highly localized behavior (Kim *et al.*, 2020).

5. *Checking Point Method (JC Method)*: The checking point method (JC method) was proposed by Rackwitz *et al.* based on the first-order second-moment theory. It has the characteristics of being able to consider non-normal random variables and having high calculation accuracy. This method has been recommended by the Joint Committee on Structural Safety (JCSS), so it is called the JC method.

The basic principle of the checking point method is that using the Taylor series to expand the function on the standard normal space at the design checking point through the process of "equivalent normalization" and calculates the random reliability after linearization. At the same time, it can linearize the nonlinear function or system equation according to the specific situation, which is convenient for analysis. Therefore, the JC method has been widely used in many fields. The MC method and JC method are used to estimate the reliability of predicted water yield of mine, which shows that the JC method is feasible and practical to calculate the random risk (Lei, 2018).

4.3.3   *Fuzzy uncertainty method*

Zadeh (1965) first proposed fuzzy set theory, which studies the relationship between input and output of a system. The detailed description of the fuzzy set theory could find in Terano *et al.* (1992), Peckhaus (2010), etc.

The fuzzy theory uses membership degree instead of probability to quantify the evaluation results, so it requires relatively low data richness and accuracy and has wider applicability than the Monte Carlo method. The triangular fuzzy number is one of the most commonly used fuzzy mathematics methods. It is mainly used to process fuzzy information with less data or low data accuracy. It has good application in groundwater vulnerability assessment, groundwater quality assessment, heavy metal ecological risk assessment, etc. There are many uncontrollable variables in the groundwater system, and it is difficult to obtain the exact values of the parameters in the survey process, i.e., there is fuzziness for the parameters. The fuzzy analysis method regards the unknown parameters as fuzzy variables, applies the fuzzy set theory to establish the membership function of the unknown parameters, converts the literal description into the mathematical description, and calculates the uncertainty of the system according to the fuzzy relation operation rules.

Dou *et al.* (1995) used fuzzy set theory and fuzzy mathematical algorithms to simulate the transport of pollutants in one-dimensional and two-dimensional homogeneous flow in aquifers and compared the differences in simulation results from fuzzy mathematical methods and results from analytical solution. Schulz *et al.* (1999) proposed a method based on fuzzy set theory, which incorporated inaccurate thermodynamic parameters into the chemical equilibrium calculation of the water system and simulated the flow in a one-dimensional unsaturated water body in the aquifer.

The establishment process of the fuzzy comprehensive evaluation model is as follows:

Given two finite groups $U = [u_1, u_2, \ldots, u_m]$, $V = [v_1, v_2, \ldots, v_n]$, U denotes a synthesis of all evaluation factors; V is the sum of the ratings, r_{ij} is the evaluated result of v_j according to evaluation factor ui, so the judgment matrix of evaluation factors (the amount is m) is (Wang *et al.*, 2005):

$$R = \begin{bmatrix} R_1 \\ R_2 \\ \vdots \\ R_m \end{bmatrix} = \begin{bmatrix} r_{11} & r_{12} & \cdots & r_{1n} \\ r_{21} & r_{22} & \cdots & r_{2n} \\ \vdots & \vdots & \vdots & \vdots \\ r_{m1} & r_{m2} & \cdots & r_{mn} \end{bmatrix} \qquad (4.11)$$

R is a fuzzy connection of V and U. If the weight of each evaluation factor is $A = [a_1, a_2, \ldots, a_m]$ (A is a fuzzy subset of U, $0 \le a_i \le 1$, the sum of a_i is 1), a fuzzy subclass of V can be gotten by applying a calculation of fuzzy transformations, i.e., the comprehensive evaluation result is:

$$B = A \cdot R = [b_1, b_2, \ldots, b_n] \qquad (4.12)$$

where B represents the fuzzy set of V, fuzzy transforms $A \cdot R$ represents the factors in multiple directions that can be applied to the universal matrix operations of multifactorial sequences. The calculation can be described as:

$$b_j = \min \left\{ 1, \sum_{i=1}^{m} a_i r_{ij} \right\} \qquad (4.13)$$

The computing sample can be found in work from Meng *et al.* (2009).

There are some deficiencies in the practical application of the fuzzy uncertainty method like the fuzzification of parameters lacks sufficient practical basis and does not consider the influence of time, space, experimental errors.

4.3.4 Grey system theory method (GST)

The grey system theory was founded in 1982 by Professor Deng Julong, and it is a research method used to study systems with a small amount of data and uncertainty. After more than 30 years of development and optimization, the grey theory has received great attention from academia all over the world. It has done well not only in hydrology but also in other fields. The essence of grey system theory is to accumulate irregular original data to generate a new sequence, and the obtained sequence has strong regularity, and then re-model. The data obtained from the generated model are then used to obtain a restored model, which is used as a prediction model (Wang *et al.*, 2002). The grey theory includes the model system based on Grey Model (GM) and the analysis system based on Grey Relational Analysis (GRA).

Grey system is between the white and black systems, only some information of it is known. The grey model takes the system with small sample size, poor information, and the irregular as the research object, which makes up for the shortcomings of classical statistical methods, to effectively extract valuable information of the system, and grasp the overall change trend and law of the system.

The specific steps to establish a gray model are as follows (Yan, 2018):

1. One-time cumulative generation sequence

Set the time series $x^{(0)}$ has n original non-negative observation $x^{(0)} = \{x^{(0)}(1), x^{(0)}(2), \ldots, x^{(0)}(n)\}$, $x^{(1)} = \{x^{(1)}(1), x^{(1)}(2), \ldots, x^{(1)}(n)\}$ is generated after accumulation, which weakens randomness and strengthens regularity.

$$x^{(1)}(k) = \sum_{i=1}^{k} x^{(0)}(i), \quad k = 1, 2, \ldots, n \tag{4.14}$$

2. A homogeneous sequence is generated by accumulated sequence

The average value is obtained from the adjacent figures to analyze the development of the system.

$$Z(k) = \frac{1}{2} \left[x^{(1)}(k) + x^{(1)}(k-1) \right] \quad k = 2, 3, \ldots, n \qquad (4.15)$$

$$Z(1) = x^{(1)}(1) \qquad (4.16)$$

3. Establishing GM (1, 1) model

The first-order linear differential equations $\frac{dx^{(1)}}{dt} + \alpha x^{(1)} = \beta$ is constructed, according to the definition of the derivative, it can be expressed as follows:

$$\frac{\Delta x}{\Delta t} = \frac{x^{(1)}(k+1) - x^{(1)}(k)}{k+1-k} = x^{(1)}(k+1) - x^{(1)}(k) = x^{(0)}(k+1)$$

$$(4.17)$$

x is taken the average value of those in time k and $k+1$, so it can be written in the following form: $Y = BA$

$$Y = ((x^{(0)}(2), x^{(0)}(3), \ldots, x^{(0)}(n))^T, \quad B = \begin{pmatrix} -Z^{(1)}(2) & 1 \\ -Z^{(1)}(3) & 1 \\ \cdots & \cdots \\ -Z^{(1)}(n) & 1 \end{pmatrix},$$

$$A = \begin{pmatrix} \alpha \\ \beta \end{pmatrix} \qquad (4.18)$$

where α is the development coefficients, reflecting the growth rate and development trend of sequence $x^{(0)}$; β is the grey action parameter, which reflects the relationship change among the data.

Least square method is used to solve the equation $\hat{A} = (B^T B)^{-1} B^T Y = \begin{pmatrix} \hat{\alpha} \\ \hat{\beta} \end{pmatrix}$, bring $\hat{\alpha}$, $\hat{\beta}$ into the original equation:

$$x^{(1)}(t) = \left[x^{(0)}(1) - \frac{\hat{\alpha}}{\hat{\beta}} \right] e^{-\hat{\alpha}k} + \frac{\hat{\beta}}{\hat{\alpha}} \qquad (4.19)$$

$$x^{(1)}(k+1) = \left[x^{(0)}(1) - \frac{\hat{\beta}}{\hat{\alpha}} \right] e^{-\hat{\alpha}k} + \frac{\hat{\beta}}{\hat{\alpha}}, k = 1, 2, \ldots, n \qquad (4.20)$$

4. Depressive restore

$$\hat{x}^{(0)}(k+1) = \left[\hat{x}^{(1)}(k+1) - \hat{x}^{(1)}(k)\right]$$

$$= (1 - e^{\hat{a}})\left(x^{(0)}(1) - \frac{\hat{\beta}}{\hat{\alpha}}\right)e^{-\hat{a}k}, \quad k = 1, 2, \ldots, n-1$$

$$(4.21)$$

The Grey model is often used to predict groundwater level, which has the advantages of less required data, high accuracy, and simple calculation. With the rapid development of computer technology, the hybrid model has gradually become the best solution to make up for the shortcomings of a single model. For example, Yan (2018) combines the grey model and machine learning method to create a mixed model, which is effectively applied to groundwater level time series prediction and spatial missing data repair in the Heihe region. Wu *et al.* (2012) combined grey theory and neural network theory to construct a grey–neural network model of groundwater prediction in the Xiaonanhai Springs area, and the predicting results is better than a single model.

4.3.5   *Unknown theory methods*

Unknown differs from randomness and ambiguity, it results from the limitations of conditions that are uninformed about the problem that has occurred.

Objectively, the uncertainty of information is not a single, but a mixture of various uncertainties. With the help of the concept of "geometric chaos", the complex information is called "information chaos". In the information chaos class, that there is two or more complicated uncertain information with randomness, fuzziness, grey, and unknown, is called blind information. Blind data is the mathematical tool to express and process blind information (Wang, 1990).

Set $G(I)$ is the grey set of grey intervals x_i, so $x_i \in G(I)$. If $\alpha_i \in [0, 1]$ $i = 1, 2, \ldots, n$, $f(x)$ is the gray function of $G(I)$. And when $i \neq j$, $x_i \neq x_j$, $\sum_{i=1}^{n}\alpha_i = \alpha \leq 1$, the function $f(x)$ is called the blind data. α_i is called the confidence level of x_i for $f(x)$ of $f(x)$, α is called the total confidence level of $f(x)$, n is called the order of $f(x)$. Here, if x_i is real, the function $f(x)$ degenerated into an unknown rational number.

From this definition, it can be seen that blind data include interval grey data and unknown rational data. Blind data is a generalization of interval data and random variable distribution. Therefore, the true blind data at least contains two kinds of uncertainties. The algorithm of blind data can be found in the relevant literature (Wang, 1990).

No matter whether the objective thing itself is certain or not, as long as it has the unknown property, the decision maker can only regard it as uncertain. Subjective membership degree or subjective probability can be used to describe the uncertainty of the thing as subjective credibility. Unknown mathematical methods have been applied to some extent, for example, Li *et al.* (2006) have conducted a preliminary study on the risk analysis of groundwater resources by unknown mathematical theory. Based on defining the basic concepts of a blind data of hydrogeological parameters and unascertained risk, they have tentatively put forward the calculation model of groundwater resources recharge under blind information and the unascertained risk analysis method of allowable exploitation.

4.3.6 *Multi-standard decision analysis (MCDA) technology based on geographic information system (GIS)*

Analyzing dynamic changes in groundwater systems require long-term survey data. The data limitation can be overcome to some extent with the help of geospatial and multi-standard decision analysis (MCDA) techniques. MCDA techniques include catastrophe theory (CT), analytic hierarchy process (AHP), and entropy theory (CT), which can be described as follows:

1. *Catastrophe theory*: René Thom, a French mathematician published the book "Structural Stability and Morphogenesis" in 1972, which marked the birth of catastrophe theory. The catastrophe theory is a mathematical method to study the phenomenon of discontinuous change of the system. Its greatest characteristic is that it can be used to explain the phenomenon of sudden "jump" without a change process. The essence of catastrophe theory is using mathematical functions to solve practical problems and regards the variables affecting the system as an energy system composed of control variables and

Table 4.4.   Primary catastrophe models.

Model types	Standard function	Number of control variables	Number of state variables
Folding	$V(x) = \frac{1}{3}x^3 + ax$	1	1
Cusp point	$V(x) = \frac{1}{4}x^4 + \frac{1}{2}ax^2 + bx$	2	1
Dovetail	$V(x) = \frac{1}{5}x^5 + \frac{1}{3}ax^3 + \frac{1}{2}bx^2 + cx$	3	1
Butterfly	$V(x) = \frac{1}{6}x^6 + \frac{1}{4}ax^4 + \frac{1}{3}bx^3 + \frac{1}{2}cx^2 + dx$	4	1
Hyperbolic umbilical point	$V(x) = x^3 + y^3 - ax - by + cxy$	3	2
Elliptical umbilical point	$V(x) = x^3 + ax^2 + bx + c(x^2 + y^2) - xy^2$	3	2
Parabolic umbilical point	$V(x) = x^4 + ax^2 + bx - x^2y + cy^2 + dy$	4	2

state variables. The change process involved in the theory is similar to the creep process, which can be understood as the process of energy and material exchange inside and outside the system.

The catastrophe theory is suitable for biological fields, economic fields, management fields, military fields, and engineering fields, it's helpful to well analyze problems and grasp the develop trend of the system. Seven primary catastrophe models have formed after long-term development (Table 4.4).

For the folding model, it is simple and easy to understand, and the fitting construction is also very easy. It does not need to master complex mathematical means. However, the accuracy of this method is insufficient and the factors considered are too few to conform to reality.

For the cusp model, it can be regarded as the evolution of the folding model, which also has the advantages of simple, easy fitting, and easy derivation. The control variables of this model are two, which will greatly improve the accuracy compared with the folding model. At present, this method has been applied most widely and maturely.

For the dovetail model, the difficulty of model construction is increased, and a certain mathematical ability is required before theoretical derivation can be made. However, the considered factors of the model are further improved, and the fitting degree and accuracy are further improved, too. For groundwater systems with a certain degree of complexity, it can be adopted. At present, this method has many applications, and many scholars will try to adopt this model when considering the influence of multiple factors.

For the butterfly model, the difficulty of the model is the highest among the single state variable models, but it also has the best accuracy, and considering the water, which is the most common influencing factor in the groundwater system. However, this analysis model requires the operator having a high mathematical ability, and the derivation of the expression and judgment basis of this model requires a variety of mathematical skills.

Hyperbolic umbilicus, elliptic umbilicus, and parabolic umbilicus are relatively advanced catastrophe models. Using these three methods requires the operator has a high mathematical knowledge and master many mathematical software at the same time, which is seldom used in the uncertainty characterization of the groundwater system.

In groundwater system, the catastrophe theory is mainly applied to groundwater environmental risk assessment, groundwater quality and vulnerability assessment, groundwater water resources carrying capacity assessment, etc (Kaur *et al.*, 2020).

2. *Analytic hierarchy process (AHP)*: In the 1970s, T.L. Saaty proposed the Analytical Hierarchy Process (AHP), which was first applied in the field of network system theory and multi-objective. The main purpose is to solve the problem that some evaluation indexes overlap in layers and each index cannot be directly quantitatively analyzed. The analytic hierarchy process of target system has great advantages by decomposing multiple interrelated influence factors into different hierarchical structures according to different attributes. The premise of using this method is that there must be certain subordinate relations between the levels.

The Analytic Hierarchy Process (AHP) is a decision-making method that decomposes the elements into objectives, criteria, schemes, and makes a qualitative and quantitative analysis on this basis. The scheme level is composed of control indexes that determine the changes of the above levels. AHP has been widely used in the field of environmental geology, such as groundwater functional zoning, engineering suitability evaluation, and environmental carrying capacity evaluation. Jenifer and Jha (2017) validated the effectiveness of hierarchical analysis, catastrophe theory methods, and maximum entropy in the evaluation of groundwater potential in hard rock aquifer systems. The results show that the hierarchical analysis is the most reliable in groundwater resource evaluations.

When modeling, AHP can be divided into six steps (Cao, 2020), namely:

(1) To determine the problems and the evaluation plan, and to evaluate the objectives;
(2) All factors in the planning system will be analyzed in depth, and then the system hierarchical structure model will be constructed;
(3) Comparing the factors of each level with the factors of the high level in pairs, and then establishing the judgment matrix of each level;
(4) Examining each matrix to judge the consistency of these matrices;
(5) To calculate the ranking of each level, and obtain the weight of the indexes of each level compared with the high level, i.e., calculate the eigenvector of the judgment matrix of each level;
(6) To calculate the total ranking of levels, check the consistency of matrices, and calculate the weight of the evaluation plan relative to the evaluation target.

The advantages of the analytic hierarchy process are simple theoretical knowledge, and the making decision process does not need any complicated calculation or all data related to factors. It has strong engineering practicability. The analytic hierarchy process combines quantitative analysis with qualitative analysis. Because it uses a relatively standard scale, it can uniformly calculate factors of different dimensions, when solving decision-making problems, the research problem is first regarded as a system, then the relationship between factors in the system and the environment is analyzed, then the management scheme of the problem is obtained.

The analytic hierarchy process also has certain defects, such as it is very difficult to check the consistency of judgment matrix. More and more groundwater characterization methods tend to combine various theories to make up for the defects of a single characterization method.

3. *Entropy theory*: The concept of entropy has experienced more than 100 years of development. In 1856, it was introduced into thermodynamics by German physicist R. Clausius. In 1866, L. Boltzman established the theory of statistical thermodynamics and put forward the entropy of statistical mechanics. In 1948, to study the uncertainty of information, Shannon called the average information amount of source signals in the communication process as entropy and took entropy as a measure to describe the uncertainty degree of the system, thus the application field of entropy concept was unprecedented expansion, which named as information entropy.

$$H = -\sum_{i=1}^{n} p_i \log\left(p_i\right) \tag{4.22}$$

where H denotes the information entropy, n represents the number of possible states of the system, and p_i is the probability of that state. In 1957, E.T. Jaynes published a milestone paper, and for the first time explicitly put forward the Principle of Maximum Entropy (POME), which successfully solved the ill-posed problem widely existing in information science. Since then, the theory of maximum entropy has been widely used in many research fields. With the development of information theory, various types of information entropy concepts have been developed to analyze system uncertainties, such as maximum entropy principle, mutual entropy, relative entropy, fuzzy entropy, entropy rate, combined entropy, and cross-entropy (Chen, 2013).

The maximum entropy principle (POME) is designed to deduce the probability distribution of random variables. Its main idea is that when only some knowledge about unknown distribution is mastered, the probability distribution that conforms to this knowledge and obtains the maximum information entropy should be selected.

Because in this case, there may be multiple probability distributions that meet the restriction conditions. Choosing the distribution with the largest information entropy means that no artificial assumptions and constraints are made on random variables, thus the characteristics of random variables can be most reasonably explained. The advantage of the maximum entropy principle is that it can eliminate the interference of human factors and risk factors and can reflect the objective information of the evaluation object.

Relative entropy, or also known as Kullback–Leibler divergence, information divergence, is used to represent the difference (or distance) between the two probability distributions (p, q). In general, p represents the true distribution of the data, q is the theoretical, analog distribution of the data, or approximate distribution of p, so that relative entropy can represent how close the approximate distribution is to the real distribution.

Mutual entropy is used to measure the interdependence of two random variables. It is defined as the reduction of uncertainty of one variable due to the existence of the other variable. Mutual entropy is a very effective sensitivity analysis method, which can analyze the complex correlation between variables. The groundwater model has many input and output variables and the model structure is complex. The relationship between model output and model variables is not simply linear or monotonous, so that the common sensitivity analysis techniques (such as stepwise regression analysis) cannot deal with the relationship well. Mutual entropy analysis provides an effective processing framework for describing the relationship between multi-variables, which can effectively identify the uncertain importance of variables.

4.3.7   *Bayes theory*

In the groundwater system, the complexity of hydrogeological parameters is a difficult problem in parameter identification. At present, the Bayesian inference method has been applied to parameter identification when pursuing a more stable and efficient global optimal solution algorithm. The parameter identification method based on Bayesian inference is a probability analysis method based on the Bayesian theorem, which can obtain the probability distribution of parameters in the whole domain, thus obtaining the global optimal solution of parameters easily. At the same time, the Bayesian method can combine the knowledge of prior information and conditional information and can give clear posterior distribution information of parameters, which is widely used in parameter uncertainty identification.

The statistician Neman highly summarized three kinds of information available in statistical inference: overall information, sample information, and prior information. As a statistical inference method, the Bayesian theorem holds that information is divided into two categories: data information and prior information. Data information is conveyed by the existing data, while prior information exist before obtained the actual data. Superimposing the data information and prior information will obtain the posterior information of the sample. Generally, Bayesian inference using posterior distribution is as follows: firstly, a distribution family of random variables for parameters is identified, then a prior distribution of parameters is determined according to experience or other information, and finally, a posterior distribution is obtained according to the prior distribution. When the Bayesian inference method is used for parameter identification, it is necessary to obtain the posterior distribution of parameters according to the prior information.

The basic form of the Bayesian theorem is as follows:

$$p(\theta|x) = \frac{p(x|\theta)p(\theta)}{p(x)} \tag{4.23}$$

$p(\theta|x)$ is the posterior probability distribution function of the parameter θ, $p(x|\theta)$ is the sampling distribution function of the parameter θ, $p(\theta)$ is the prior distribution function of the parameter θ, and $p(x)$ is the edge distribution function of x.

When the Bayesian inference method is applied to parameter identification, three types of sample information, namely, the overall information of the sample, the prior information, and the sampling information, are combined at the same time. Because the generation of parameters in this method is random, large number of sampling is needed to get the representative sample points when there are many parameters, and it will increase the computation.

Chapter 5

Research Examples

5.1 Research Cases Based on Theoretical Analysis of Complexity Science

5.1.1 *Application of dissipative structure theory to analyze of groundwater system evolution, taking Hengshui City as an example*

This study was carried out by Jiang *et al.* (2008). The conclusions of dissipative structure theory on the study of systems, such as non-equilibrium, openness, and orderliness of systems, and nonlinear effects, have made outstanding contributions to the theory of nonlinear systems (Zhu *et al.*, 2002), which has laid a rigorous scientific foundation for the further development of systems science (Cheng *et al.*, 2003). The process of groundwater systems evolution is that old macrostates replaced by new macrostates under the influence of natural or human activities. The evolution of the groundwater system consists of a causal chain, which shows the transformation from a stable state to an unstable state and then from an unstable state to a new stable state (Gao and Dong, 2003). All spontaneous processes in nature are irreversible, so the states experienced in the evolution process are all non-equilibrium states, and non-equilibrium structures can be formed in the evolution process of non-equilibrium states (Li *et al.*, 1996). The water resources system is a dissipative structure (Ding and Li, 2001), and the evolution process of the water cycle is dissipative (Zheng and Gao, 2000). The evolution of the groundwater system is the evolution process of a nonlinear and complex system. There is extensive communication of matter, energy, and information between the groundwater system and the outside world. It is a complex and open system that is neither isolated nor closed. Prigogine (1964) believed that when a sustained exchange of energy and mass occurred between the open system and the environment, it would be possible for the system to be moved away from the equilibrium state and that the consumption of system energy due to irreversible processes could make it "self-organized" and generate an orderly "dissipative structure" temporal and spatial. The large-scale exploitation of groundwater since 1970 have had a great impact on the groundwater environment in Hengshui City, that destroyed the natural water circulation, and changed the recharge of groundwater resources, causing significant changes in the quantity, quality, and temporal and spatial distribution of groundwater resources. It is of great theoretical and practical significance to discuss the evolution mechanism and trend of groundwater systems under the influence of human activities. The study is based on dissipative structures to explore the evolutionary mechanisms of groundwater systems under the influence of manmade activities, to provide a new research idea for groundwater system analysis.

Since the 1970s, under the strong interference of human activities, the flow field, spatial structure of groundwater medium, and hydrochemical field of Hengshui groundwater system had successively entered abnormal fluctuation states, and the groundwater system had evolved toward a new state. Before 1970, the groundwater system in the Hengshui area was basically in a macro-stable state, and the medium spatial structure, flow field, and hydrochemical field of groundwater were in a normal fluctuation state without obvious trend changes. After 1970, with the increase of groundwater exploitation and the deepening of exploitation depth, the groundwater level fluctuated abnormally, the water level continued to decline, and the fluctuation range gradually increased. The groundwater system entered an unstable state due to imbalance. The groundwater level and water quality fluctuated abnormally, forming a large area of water cone of depression ("Ji Zao Heng" depression). The salinity of groundwater increased and the hydrochemical type changed. At the same time, the medium structure of groundwater had also begun to evolve, resulting in geological disasters such as land subsidence.

5.1.1.1 *Evolution of groundwater flow field*

The quality of deep groundwater in Hengshui City was good, and the total dissolved solids (TDS) was less than 1 g/L. Since 1968, the third water-bearing group had been exploited on a large scale for a long time. The exploitable amount of deep groundwater is 243 million m^3/a, the actual production since 1992 had exceeded 500 million m^3/a and is increasing year by year. The exploitation of deep groundwater far exceeded its recharge, which led to the continuous decline of groundwater level, variation of the groundwater flow field. The macro-stable state of groundwater flow system was destructed, and the flow field controlled by artificial sources and sinks formed.

The "Ji Zao Heng" deep depression was formed in the early 1970s. At the beginning of 1968, the water buried depth of the third water-bearing formation was only 2.94 m. With the rapid increase of deep groundwater exploitation, the water level decreased continuously. In 1972, the buried depth of the water in the depression center increased to 20.98 m, with an annual average decrease of 4.51 m. In 1990, the maximum depth of the water level in the center of the depression was 56.4 m, the water level in the center of the depression dropped to 2.24 m/a, and the depression area reached 4032 km^2, close to half of Hengshui's total area. In 1993, the depression began to gradually spread out of the area. The eastern drainage divide swung left and right due to the influence of depression. At the end of June 2000, the buried depth of the water in the depression center reached 101.00 m, which was 19.10 m lower than that of the same period in 1995, and the average reduction speed was 3.82 m/a, as shown in Figure 5.1.

The depth of deep groundwater level in Zaoqiang and the south of Wuyi, which were located in the middle of the depression area, was close to that in Hengshui City, and the depth of water level in the south of Jizhou had exceeded that in Hengshui City. The difference between the water level in the center of the depression and the eastern drainage divide gradually decreased, and the bottom of the depression became more open and flat, showing a trend in depth, as shown in Figure 5.2.

5.1.1.2 *Spatial structure evolution of groundwater medium*

Since the 1970s, with the large-scale exploitation of groundwater, while the groundwater flow field fluctuated abnormally, the medium

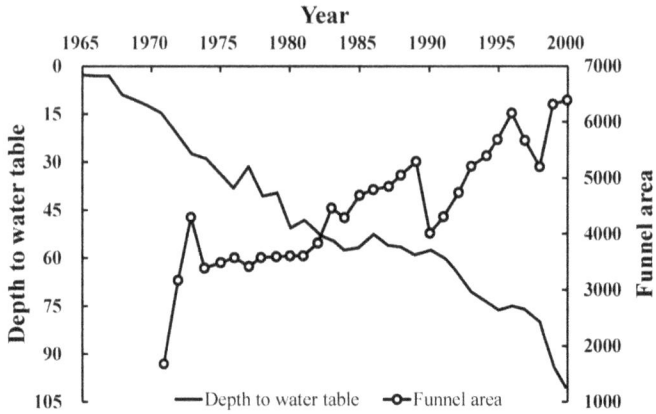

Figure 5.1.   "Ji Zao Heng" funnel dynamic feature curve (Jiang *et al.*, 2008).

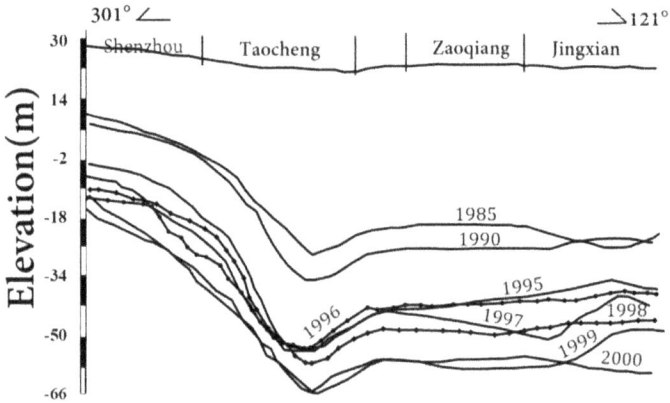

Figure 5.2.   "Ji Zao Heng" funnel groundwater level dynamic profile (Jiang *et al.*, 2008).

spatial structure of the groundwater system also showed abnormal variation, resulting in geological disasters such as land subsidence and ground fissures. In 1975–1981, the "Ji Zao Heng" depression zone began to appear land subsidence. From 1981 to 1988, with the instability of groundwater system, the range and rate of land subsidence began to increase, with the subsidence amount reaching 128 mm and the subsidence rate of 16 mm/a at this stage. With the acceleration of the decline rate of groundwater level, the rate of land subsidence accelerated and the area further expanded. From 1988 to 1990, the subsidence was 51 mm and the rate was 25.5 mm/a. The largest subsidence was in the center of Hengshui city, which gradually decreased from the center to the periphery of the depression. The distribution characteristics of land subsidence were consistent with the shape of the groundwater depression cone.

5.1.1.3 *Evolution of groundwater chemical field*

The evolution of the groundwater chemical field lags behind that of the flow field, but the integrity of the groundwater system determines that the groundwater chemical field should eventually adapt to the evolution process of the groundwater flow field. In areas with strong human activities, groundwater quality was abnormal, mainly manifested by salinization of freshwater, as shown in Figure 5.3.

The "Ji Zao Heng" depression became an artificial sink area. Groundwater converged toward the center of the depression, and the sink of the hydrochemical field also tended to the center of the depression. Also, the compaction of clayey soil accelerated the interaction between the solid phase and liquid phase, causing ions in the solid phase to release, resulting in a slow increase in the mineralization of groundwater in the depression area, thus forming a trend of increasing the mineralization of groundwater from the periphery of

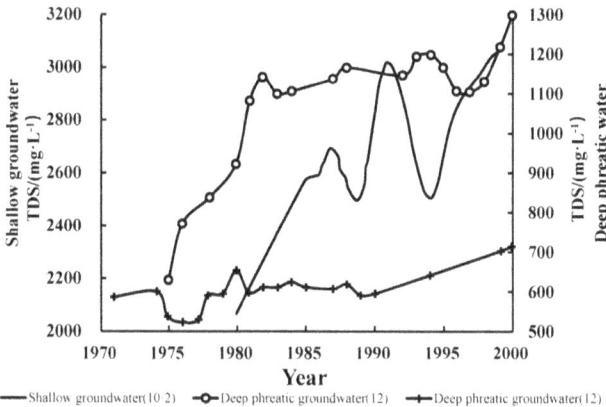

Figure 5.3. Anomaly of water quality in the groundwater system (Jiang *et al.*, 2008).

the depression to the center. Except for Anping and Raoyang counties and some areas of Shenzhou city which were all freshwater areas, salty water bodies with different thicknesses generally exist in the upper part of deep underground fresh water in other counties of the city. The large drop of deep groundwater level led to the increase of water head difference between the shallow aquifer and deep aquifer. The large hydraulic gradient drove the shallow groundwater to vertically overflow and recharge the deep groundwater. The upper saline water gradually moved down, with a moving rate of 0.1–0.2 m/a, which made the deep groundwater deteriorate.

Under mining conditions, the water level in the freshwater area declined, causing saltwater to intrude into the freshwater area, resulting in increasing of freshwater salinity. At the same time, the decrease of phreatic water level resulted in the decrease of evaporation, the downward movement of soil salt, and the enhancement of leaching, which made a large amount of salt enter groundwater, so that the shallow groundwater quality changed.

5.1.1.4 *Conclusion*

(1) The groundwater system is a system with a dissipative structure. The groundwater system is a complex open system, and its evolution process is in a non-equilibrium ordered state and irreversible. The interaction between its related subsystems is nonlinear. Strong human factors have a significant impact on its evolution. The evolution of the groundwater system goes through a process of the macro-stable state, unstable state, and reconstructed stable state.

(2) The "Ji Zao Heng" depression is the product of groundwater system evolution under strong interference from human activities. Large-scale overexploitation of groundwater leads to the variation of groundwater dynamic field and chemical field in Hengshui City. Human activities make the groundwater system enter a new evolution period with the strong interference of human factors.

(3) At present, human beings cannot completely shape the new groundwater system according to their wishes without causing other adverse environmental problems. Therefore, system instability or functional changes are often one of the constraints that must be avoided when utilizing groundwater.

5.1.2 Complexity analysis of groundwater buried depth sequence based on sample entropy

This study was carried out by Liu and Liu (2012). Jiansanjiang Sub-bureau is located in the northeast of Heilongjiang Province and the hinterland of Sanjiang Plain. It mainly develops rice cultivation and is an important grain reserve base and commodity grain production base in China. In recent years, influenced by the national grain policy and stimulated by economic benefits, the area of paddy fields has increased greatly, resulting in a significant increase in the water resources needed for irrigation, causing problems such as overexploitation of groundwater, irrational allocation of water resources, and sudden change in the water resources utilization. As a result, the complexity of the groundwater resources system is getting worse, which has attracted the general attention of domestic scholars. However, previous studies on water resources systems often ignored its complexity, which made it difficult to truly reveal the essence of the water resources system. Therefore, an in-depth study on the complexity of groundwater buried depth sequence is the key to realize scientific utilization and management of water resources.

5.1.2.1 Basic principle of sample entropy

Sample entropy is a statistic that is different from approximate entropy and does not include its matching. It represents the probability that nonlinear dynamic systems generate new information and is mainly used to quantitatively describe the regularity and complexity of the system. The larger the sample entropy value, the lower the sequence's self-similarity, the higher the probability of generating new information, and the more complex the sequence (Peng *et al.*, 2009). SampEn (m, r, N) is used to represent the sample entropy and its specific algorithm (Ge *et al.*, 2008) as follows:

Let the original data be $x(1), x(2), \ldots, x(N)$, with N points.

(1) Compose a set of m-dimensional vectors in sequential order, from $X_m(1)$ to $X_m(N - m + 1)$, where

$$X_m i = [x(i), x(i+1), \ldots, x(i+m-1)], \quad i = 1 \sim N - m + 1$$

$$(5.1)$$

Define the distance $d[X_m(i), X_m(j)]$ between $X(i)$ and $X(j)$ as the one with the largest difference of the corresponding elements, i.e.,

$$d[X_m(i), X_m(j)] = \max[x(i+k) - x(j+k)], \quad k = 0 \sim m - 1;$$

$$i, j = 1 \sim N - m + 1; \quad i \neq j \quad (5.2)$$

(2) Given that threshold r, for each value $i(1 \leq i \leq N - m)$, counting the number (the number of template matches) that $d[X_m(i), X_m(j)]$ is less than r and the ratio of this number to the total number of distances $(N - m - 1)$, denoted as $B_i^m(r)$

$$B_i^m(r) = \{number\ of\ (d[X_m(i), X_m(j)] < r)\}/(N - m - 1),$$

$$j = 1 \sim N - m; \quad i \neq j \quad (5.3)$$

The average value is obtained for all i, that is,

$$B^m(r) = \left\{ \frac{1}{(N-m)} \right\} \times \sum_{i}^{N-m} B_i^m(r) \quad (5.4)$$

The dimension is increased by 1 to become an $m + 1$ dimension vector, and the steps of (1) to (3) are repeated to obtain

$$B^{m+1}(r) = \left\{ \frac{1}{(N-m)} \right\} \times \sum_{i}^{N-m} B_i^{m+1}(r) \quad (5.5)$$

(3) $B^m(r)$ and $B^{m+1}(r)$ are similar probabilities of the sequence of m points and $m+1$ points respectively, then the theoretical sample entropy in sequence is:

$$SampEn(m, r) = \lim_{N \to \infty} \left\{ -\ln \left[\frac{B^{m+1}(r)}{B^m(r)} \right] \right\} \quad (5.6)$$

When N is finite, a sequence sample entropy estimate is derived,

$$SampEn(m, r, N) = -\ln[Bm + 1(r)/Bm(r))] \quad (5.7)$$

The selection of parameters m and r is the key to sample entropy estimation, but so far there is no optimal standard. According to previous studies (Bai *et al.*, 2007; Su *et al.*, 2011; Richman and Moorman, 2000), it is usually taking $m = 2$, $r = (0.10\text{--}0.25)$ SD, SD is the standard deviation of the original sequence.

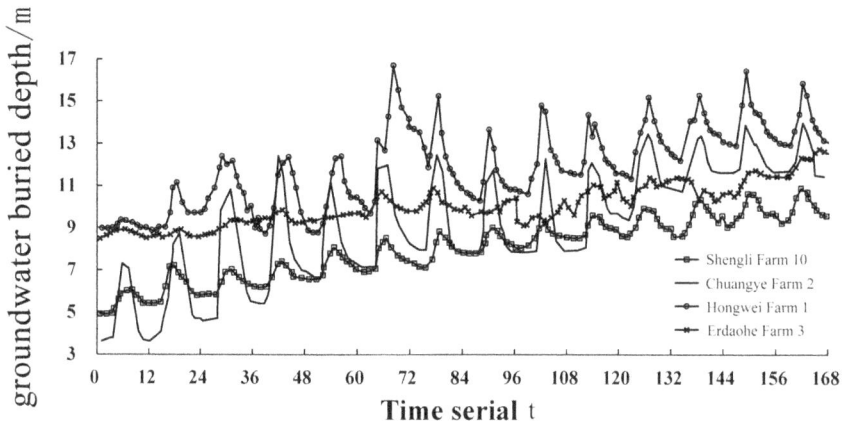

Figure 5.4. Dynamic change curve of groundwater buried depth in long monitoring wells (Liu and Liu, 2012).

5.1.2.2 *Source of data sequence*

Month-by-month groundwater level monitoring data were collected by the Water Bureau of Jiansanjiang Sub-bureau including Shengli Farm, Chuangye Farm, Hongwei Farm, Honghe Farm, and Erdaohe Farms for 1997–2010, according to which the dynamic change curve of the underground water depth sequence $H_t(t = 1, 2, 3, \ldots, 168)$ was mapped, see Figure 5.4. Only some regions including Shengli Farm 10, Chuangye Farm 2, Hongwei Farm 1 and Erdaohe Farm 3 were selected considering representativeness.

5.1.2.3 *Sample entropy parameter selection*

Taking $m = 2, k = 0.10 - 0.25$, step size of 0.01, MatlabR2009a software was taking for programming calculation, the results were gotten and shown in Table 5.1.

To determine the k-value, it is needed to determine how much the team's SampEn values change with k-values so that the k-value at the minimum change is the final k-value (Yan and Gao, 2007), see the results are shown in Figure 5.5. Only some regions including Shengli Farm 10, Chuangye Farm 2, Hongwei Farm 1 and Erdaohe Farm 3 were selected considering representativeness. It can be seen from Figure 5.5 that when k changes from 0.23 to 0.24, the average change of SampEn is only 2.02%, which has reached the minimum value. So it was taking $k = 0.24$, $r = 0.24$ SD.

Table 5.1. SampEn values for month-by-month groundwater buried depth sequence under different *k*-value conditions (Liu and Liu, 2012).

Long monitoring location	SampEn(2, kSD)								
	$k = 0.10$	$k = 0.12$	$k = 0.14$	$k = 0.16$	$k = 0.18$	$k = 0.20$	$k = 0.22$	$k = 0.24$	$k = 0.25$
Shengli Farm 31	0.8205	0.7588	0.6655	0.6162	0.5810	0.5294	0.4716	0.4474	0.4233
Shengli Farm 10	0.9389	0.8020	0.7545	0.7150	0.6654	0.6020	0.5919	0.5546	0.5091
Chuangye Farm 2	0.5383	0.5317	0.5592	0.5852	0.5798	0.5972	0.6052	0.5943	0.5841
Hongwei Farm 1	0.7632	0.7042	0.6945	0.6767	0.6472	0.5977	0.5608	0.5664	0.5694
Hongwei Farm 12	0.8676	0.7989	0.7936	0.7395	0.6672	0.6542	0.6210	0.6073	0.5969
Honghe Farm 2	0.8171	0.7332	0.6942	0.6313	0.6138	0.5589	0.5411	0.4910	0.4717
Honghe Farm 8	0.8044	0.6811	0.6152	0.5706	0.5229	0.4720	0.4470	0.4307	0.4153
Erdaohe Farm 3	0.7497	0.6631	0.5871	0.5006	0.4662	0.4409	0.4054	0.3903	0.3747
Erdaohe Farm 6	0.3286	0.2725	0.2270	0.2054	0.1973	0.1779	0.1629	0.1465	0.1354

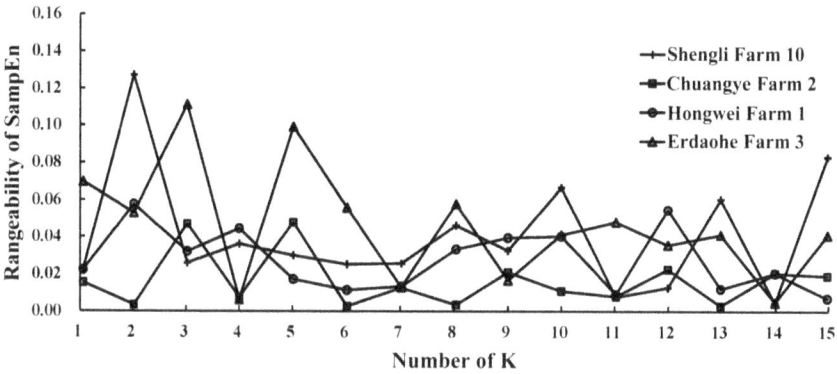

Figure 5.5. Variation curve of SampEn value with k-value in each monthly groundwater buried depth sequence (Liu and Liu, 2012).

5.1.2.4 *Results and discussion*

SampEn (2, 0.24 *SD*) was used to sort the complexity of the monthly groundwater buried depth series of the nine farms. As seen in Table 5.1, Hongwei Farm 12 > Chuangye Farm 2 > Hongwei Farm 1 > Shengli Farm 10 > Honghe Farm 2 > Shengli Farm 31 > Honghe Farm 8 > Erdaohe Farm 3 > Erdaohe Farm 6. It can be seen from this that the complexity of the monthly groundwater buried depth series of the nine teams has obvious spatial differences, and the overall performance is gradually increasing from east to west, which shows that there are more factors and uncertain conditions affecting the groundwater buried depth series in the west, so that the groundwater system structures is more complicated. The possible reasons for the difference in the overall complexity of the nine teams' groundwater depth sequences include hydrogeological conditions such as evapo-ration, precipitation, air temperature, underlying surface conditions, soil water storage conditions, and human activities such as exploit-ing groundwater for agricultural irrigation, increasing crop planting area, building drainage projects, and building reservoirs. The Sam-pEn values of Hongwei Farm 12, Chuangye Farm 2, Hongwei Farm 1, and Shengli Farm 10 are relatively high. According to the nature of sample entropy, the larger the SampEn values, the higher the com-plexity of the system, and the lower the predictability. According to relevant information, Hongwei Farm, Chuangye Farm, and Shengli Farm have paddy fields of 35333.3 hm^2, 30000 hm^2, and 26666.7

hm^2, respectively. They account for 91.38%, 93.75%, and 88.89% of the cultivated land area of their respective farms, and have a gradually increasing trend. The large proportion of paddy field area in the cultivated land area determines that these farms need to exploit a large amount of groundwater for rice irrigation. Therefore, the four production teams have a large amount of groundwater exploitation, resulting in the complexity of groundwater buried depth sequence at the top. The complexity of groundwater buried depth sequences in the two areas of Honghe Farm is relatively low, and the amount of groundwater exploitation is not large, which is further favored by evaporation, precipitation, temperature and other conditions, resulting in the complexity of groundwater buried depth sequences is normal in the second and eighth areas of Honghe Farm. The complexity of groundwater buried depth sequence in Erdaohe Farm 3 and 6 is the lowest, which is related to Wusuli River Irrigation Area Project of Bajiuwu Farm, the largest water-lifting irrigation project in Heilongjiang Province (Dong *et al.*, 2007), because Erdaohe Farm is close to Bajiuwu Farm, with the help of surface water irrigation, the underground water source is conserved to a certain extent, and Erdaohe Farm is close to Wusuli River, so the groundwater supply is sufficient. Therefore, the complexity sequence of groundwater buried depth in the third and sixth areas of Erdaohe Farm is at the end.

From the above analysis, it can be seen that human production activities are the main reasons for the complexity of the regional groundwater buried depth sequence and spatial differences. The calculation results are consistent with the actual situation and accurately reflect the use of groundwater resources in the region.

5.1.3   *Complexity diagnosis of regional groundwater resources system based on fractal dimension of time series and artificial fish swarm algorithm*

This study was carried out by Yu *et al.* (2013). In recent years, with the in-depth study of hydrological science, the analysis of the complex characteristics of regional hydrological systems has gradually become a hot topic in hydrological research (Wallis and Ison, 2011; Winz *et al.*, 2009). As an important part of the hydrological system, the groundwater resources system has typical complex characteristics.

Due to the high speed of economic development and population growth, the groundwater resources in the Jiansanjiang area in China are affected by manmade and natural factors. To realize the optimal local allocation of water resources and promote the development of the local economy, it is very important to select an appropriate complexity measurement method for groundwater resources system.

In this work, the intelligent and efficient fitting data combined with the fractal dimension of time series calculated based on curve length and the artificial fish algorithm was applied to the diagnosis of Jiansanjiang groundwater. The fractal dimension and average complexity of groundwater buried depth series in 15 districts of Jiansanjiang Sub-bureau were calculated. The results show that the complexity in the north district is the highest and that in the south district is the lowest. The most important influencing factor of local groundwater buried depth dynamics is human activities. The research shows that it is feasible to extract the complexity characteristics of hydrological time series by combining fractal theory and artificial fish algorithm, which can be applied to the study of regional hydrological processes and provide a scientific basis for the sustainable utilization of regional groundwater resources.

5.1.3.1 *Research area overview*

Jiansanjiang Sub-bureau belongs to Sanjiang Plain of Heilongjiang Province (Figure 5.6). This area is located in the northeast of China, with a total land area of 12,400 square kilometers (Liu *et al.*, 2011) and cultivated area of 68,2000 hm^2. The elevation of this area shows a downward trend from southwest to northeast, with few hills in the north and southeast. Rice is mainly grown in this area, including 15 large- and medium-sized state-owned farms. The annual average temperature in this area is between 1.0 and 2.0°C, and the highest extreme temperature is 38–41.6°C. Average annual rainfall varies.

5.1.3.2 *Fractal dimension of time series*

In recent years, the concept of fractal has been applied in many fields of natural science. The fractal theory is applied not only to topological objects but also to geometric feature rules and time series. In different steps, the time series is measured on the same time scale t.

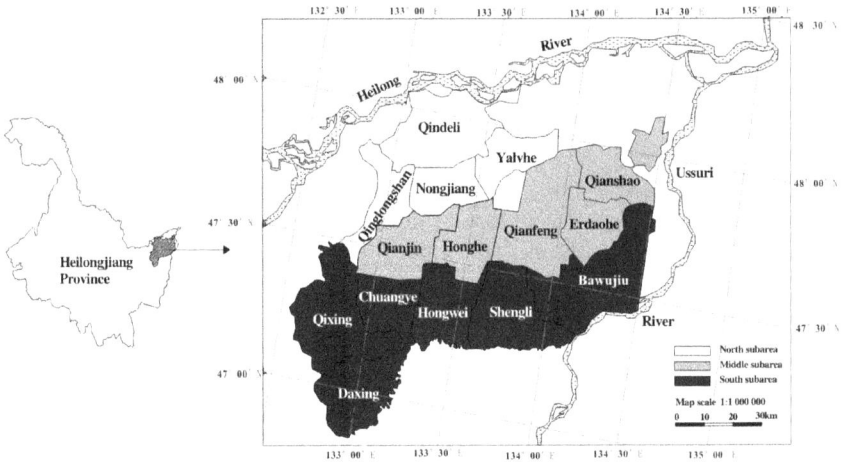

Figure 5.6.　Jiansanjiang Sub-bureau Location, Heilongjiang Province (Yu *et al.*, 2013).

When new nodes are added to the curve, the time series has scale invariance (Suleymanov *et al.*, 2009a).

The reliability of the fractal time series of dimension d is closely related to the finiteness of measurement steps. For long-term sequence research, it is necessary to measure a large amount of data and analyze the changes in dynamic trends in the measurement process (Dubovikov *et al.*, 2004). Therefore, the fast convergence of the algorithm is very important for the calculation of fractal dimension, which is limited by the number of time series measurements.

5.1.3.3　*Artificial fish fitting algorithm*

The Artificial Fish Group Algorithm (AFSA) (Figure 5.7) was first proposed in 2002 (Li and Qian, 2003) and is developing globally. It originates from the process of fish stocks searching for food. It introduces the principle of artificial intelligence based on animal behavior to solve and optimize problems through animal methods. By simulating the actual movement of fish, AFSA has created artificial fish to change their position by foraging, gathering, and tracking sargassum. After a certain period, all the artificial fish gather together. According to the situation of artificial fish, we can find the optimal value.

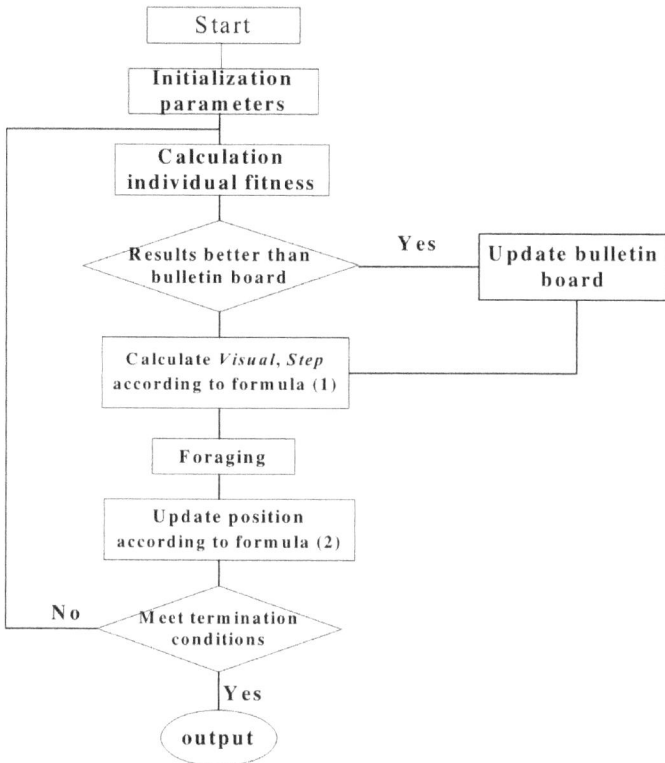

Figure 5.7.   Basic flow chart of artificial fish swarm fitting algorithm.

5.1.3.4   *Results and discussion*

5.1.3.4.1   Data source

Monthly groundwater buried depth monitoring data $(n = 168)$ of 15 districts from 1997 to 2010 were collected from the Water Supplies Bureau of Jiansanjiang Sub-bureau in Heilongjiang Province of China. The variation curve of the groundwater buried depth sequence for each farm is given in Figure 5.8. The results show that all groundwater buried depth of monitoring points in Jiansanjiang Sub-bureau show non-stationary and random variation characteristics, but the overall trend is gradually increasing. It is worth noting that the elevations of Qindeli Farm 7 and Qianshao Farm 22 are relatively high, resulting in high groundwater buried depth.

Table 5.2. Numerical results and regression equation of complexity dimension of groundwater depth series in Jiansanjiang Sub-bureau (Yu et al., 2013).

Well location for long-term monitoring	Well number for long-term monitoring	AFS fitting Regression equation	R^2	Least square method Regression equation	R^2	Fractal dimension D
Bawrjiu Farm 1	OW1	$L=0.263K^{0.027}$	0.9358	$L=0.655k^{0.428}$	0.4294	1.027
Chuangye Farm 2	OW2	$L=0.251k^{0.015}$	0.5424	$L=0.736k^{0.223}$	0.5414	1.015
Shengli Farm 31	OW3	$L=0.398k^{0.855}$	0.8736	$L=0.739k^{0.821}$	0.5464	1.855
Qindeli Farm 7	OW4	$L=0.351k^{0.422}$	0.9015	$L=0.765k^{0.317}$	0.5852	1.422
Hongwei Farm 21	OW5	$L=0.297k^{0.292}$	0.7233	$L=0.718k^{0.953}$	0.5155	1.292
Honghe Farm 6	OW6	$L=0.312k^{0.332}$	0.7731	$L=0.733k^{0.027}$	0.5373	1.332
Erdaohe Farm 5	OW7	$L=0.343k^{0.917}$	0.8049	$L=0.579k^{0.917}$	0.336	1.917
Qinglongshan Farm 17	OW8	$L=0.303k^{0.871}$	0.8937	$L=0.756k^{0.615}$	0.5716	1.871
Yalvhe Farm 5	OW9	$L=0.255k^{0.775}$	0.7998	$L=0.71k^{0.573}$	0.5047	1.775
Nongjiang Farm 8	OW10	$L=0.33k^{0.954}$	0.8379	$L=0.809k^{0.291}$	0.6552	1.954
Ministry of Qianfeng Farm	OW11	$L=0.294k^{0.41}$	0.9297	$L=0.731k^{0.409}$	0.5348	1.41
Daxing Farm 11	OW12	$L=0.287k^{0.079}$	0.8661	$L=0.857k^{0.065}$	0.7337	1.079
Qianshao Farm 12	OW13	$L=0.283k^{0.602}$	0.9344	$L=0.757k^{0.775}$	0.5735	1.602
Qixing Farm 69	OW14	$L=0.275k^{0.208}$	0.7688	$L=0.841k^{0.35}$	0.7065	1.208
Qianjin Farm 22	OW15	$L=0.365k^{0.611}$	0.8521	$L=0.862k^{0.547}$	0.7431	1.611

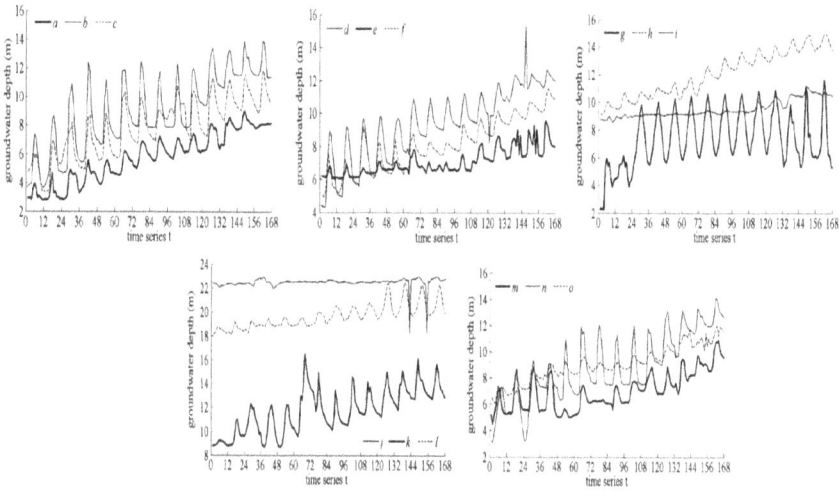

Figure 5.8. Variation curve of monthly groundwater buried depth series of each farm in Jiansanjiang Sub-bureau from 1997 to 2010. (a) Bawujiu Farm 1; (b) Chuangye Farm 2; (c) Shengli Farm 31; (d) Qindeli Farm 7; (e) Hongwei Farm 21; (f) Honghe Farm 6; (g) Erdaohe Farm 5; (h) Qinglongshan Farm 17; (i) Yalvhe Farm 5; (j) Nongjiang Farm 8; (k) Ministry of Qianfeng Farm; (l) Daxing Farm 11; (m) Qianshao Farm 22; (n) Qixing Farm 69; (o) Qianjin Farm 4 (Yu *et al.*, 2013).

5.1.3.4.2   Complexity measurement of groundwater depth series in Jiansanjiang Branch of China

As shown in the figure (Figure 5.8), the time series fractal dimension method is used to calculate the dimension values of each region. At the same time, the least square regression method and the artificial fish swarm algorithm are applied to fit the $L \sim Ak^{D-1}$ relationship, respectively, and the corresponding results are listed in Table 5.2. Comparing the results of the artificial fish swarm algorithm and the least square method shown in Table 5.2, it can be seen that the results of the first method are more accurate than those of the second method. In the regression equation obtained by the artificial fish swarm algorithm, the coefficient of each farm is lower than 0.4, and it is lower 0.3 in the Bawujiu Farm 1, Chuangye Farm 2, Hongwei Farm 21, Yalvhe Farm 5, Ministry of Qianfeng Farm, Daxing Farm 11, Qianshao Farm 12, Qixing Farm 69, which means that the fitting curve of artificial fish fitting method is closer to the real curve, and the fitting results are efficient. Also, the correlation coefficients (R^2)

between L and K revealed that the fitting result of artificial fish fitting method are more accurate because of high correlation coefficients.

Fractal dimension is the core of fractal theory and an important index to measure system complexity. The larger the fractal dimension value, the more complex the system (Zhang, 2009). As can be seen from the results in Table 5.2, the complexity of groundwater depth in Jiansanjiang Sub-bureau are as follows: Nongjiang Farm 8 > Erdaohe Farm 5 > Qinglongshan Farm 17 > Shengli Farm 31 > Yalvhe Farm 5 > Qianjin Farm 4 > Qianshao Farm 22 > Qindeli Farm 7 > Ministry of Qianfeng Farm > Honghe Farm 6 > Hongwei Farm 21 > Qixing Farm 69 > Daxing Farm 11 > Bawujiu Farm 1 > Chuangye Farm 2.

5.1.3.4.3   Analysis of the spatial distribution of underground
            water depth sequence dimension of Jiansanjiang
            Sub-bureau

Spatial map based on dimensions was obtained by GIS software and shown as Figure 5.9. Dimensions obtained from Table 5.2 are divided into five degrees. The fractal value of groundwater depth complexity in Nongjiang Farm 8, Erdaohe Farm 5, Qinglongshan Farm 17, Qianjin Farm 4 is at the highest level, indicated that groundwater depth in these areas is affected by more factors than that of other areas, and the structure of the dynamic complexity of the groundwater system is strong. On the contrary, the fractal values of groundwater depth complexity in Daxing Farm 11 and Zhuangye Farm 2 are at the lowest level.

The spatial distribution trend of the complexity of groundwater depth sequence in Jiansanjiang Sub-bureau are further measured, and the average D values of the south, central, and north areas are calculated (Table 5.3). The results show that the general trend of

Table 5.3.   Average complexity of monthly groundwater depth series in each division of Jiansanjiang Sub-bureau.

Subarea	Contained farms	Average D
North subarea	Qinglongshan, Qindeli, Nongjiang, and Yalvhe	1.756
Central subarea	Qinjin, Honghe, Qianfeng, Erdaohe, and Qianshao	1.574
South subarea	Qixing, Chuangye, Daxing, Hongwei, Shengli, and Bawujiu	1.246

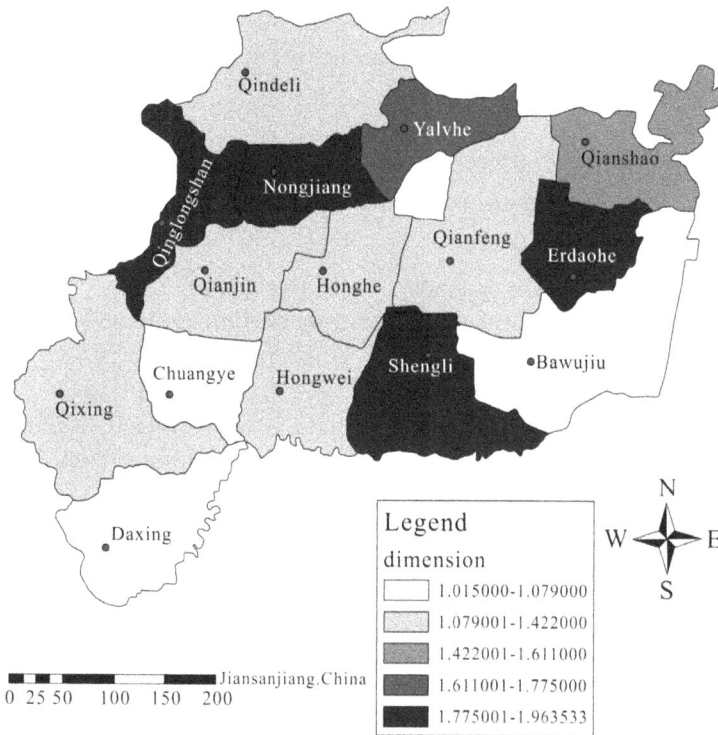

Figure 5.9. Space map of Jiansanjiang Sub-bureau dimension values results (Yu *et al.*, 2013).

complexity gradually weakens from north to south. There are many reasons for this result like the surface water resources cannot be effectively utilized in the mountainous area of Qindelhi. Qinglongshan area is the most important paddy field in Sanjiang Plain, although there are many rivers, such as Nongjiang River and Yalv River, their distribution is unreasonable and the surface water resources are not evenly distributed, which makes it difficult to effectively utilize. So why groundwater exploitation is serious and groundwater depth sequence is the most complex in the northern division. The central region is dominated by plain, the utilization of surface water is better than that of the northern region, and the level of groundwater exploitation is weaker. Therefore, the complexity of groundwater depth sequence is in the middle. In the southern division, there are irrigation projects of Qixing River, Daxing River, and

Ussuri River, which is the largest pumping irrigation project in Heilongjiang Province. Thus, the groundwater depth series complexity is the weakest in the south.

5.1.3.4.4   Analysis of factors affecting the complexity of groundwater depth sequence in Jiansanjiang Sub-bureau

According to the data of paddy field area (1997–2010) collected by groundwater depth monitoring stations of Jiansanjiang Sub-bureau, the variance is calculated (Table 5.4). Figure 5.10 shows that Yalvhe Farm 5, located in the northern division of Jiansanjiang, has the highest change in 14 years paddy field area variation, followed by Qinglongshan Farm 17, Erdaohe Farm 5, and Nongjiang Farm 8. The variance calculation results of paddy field area obtained from Table 5.4 also show that the variance of Erdaohe Farm 5, Qinglong-shan Farm 17, Yalvhe Farm 5, and Nongjiang Farm 8 is higher than

Table 5.4.   Variation of annual rainfall and paddy field area of Jiansanjiang Sub-bureau (1997–2010) (Yu *et al.*, 2013).

Well location for long-term monitoring	Paddy field area		Annual rainfall	
	Variance (10^4)	Variance contribution (%)	Variance (10^3)	Variance contribution (%)
Bawujiu Farm 1	11.0	1.38	9.49	8.28
Chuangye Farm 2	8.26	1.04	4.36	3.81
Shengli Farm 31	9.85	1.24	9.74	8.5
Qindeli Farm 7	2.38	0.3	7.70	6.72
Hongwei Farm 21	35.9	4.52	10.4	9.07
Honghe Farm 6	35.9	4.52	10.4	9.07
Erdaohe Farm 5	122.0	15.4	6.60	5.76
Qinglongshan Farm 17	153.0	19.3	9.75	8.51
Yalvhe Farm 5	269.0	34	4.09	3.57
Nongjiang Farm 8	56.9	7.17	1.88	16.4
Ministry of Qianfeng Farm	36.2	4.56	8.45	7.37
Daxing Farm 11	9.65	1.22	5.33	4.65
Qianshao Farm 12	35.9	4.52	6.35	5.54
Qixing Farm 69	19.0	2.4	2.96	2.58
Qianjin Farm 22	18.6	2.34	6.96	6.07
Total variance	**793**		**115**	

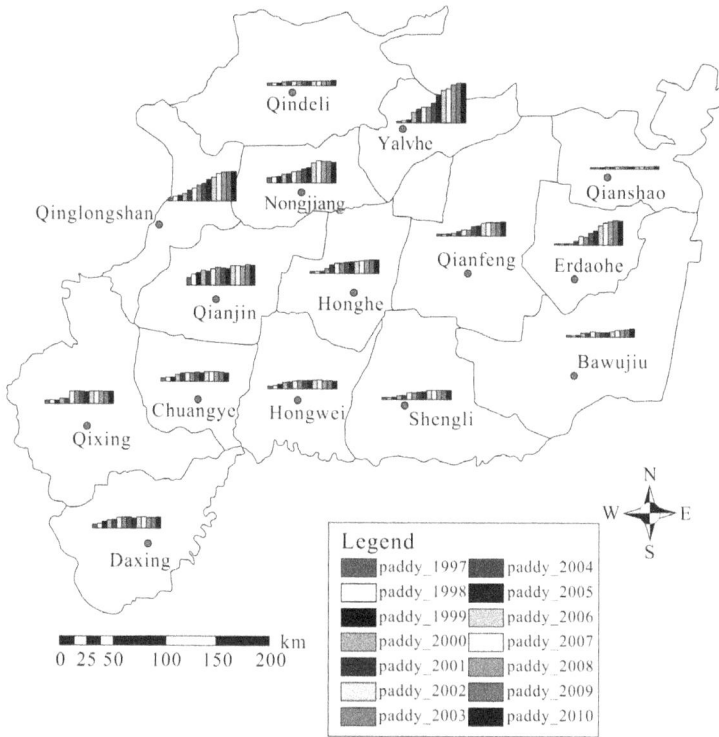

Figure 5.10. Spatial distribution map of paddy field planting area in Jiansan-jiang Sub-bureau (1997–2010) (Yu *et al.*, 2013).

that of other areas. At the same time, the values of the four districts contribute the most to the variance of population differences and have the greatest impact on the overall wet area changes of Jiansan-jiang Sub-bureau. This result is consistent with the highest dimensional value of the groundwater system complexity in the northern subregion.

The change of paddy field area in central areas of Jiansan-jiang Sub-bureau is moderate, which may be due to the combination of local climate influence and groundwater recharge conditions. This may lead to the middle complexity of groundwater depth sequence in this region, while the southern region has the lowest change for the same reason as the central region. The Jiansanjiang Sub-bureau is dominated by plain or wetland with fewer hills. The plain covers an area of 11,291 square kilometers, accounting for 92.6%

of the total area, and is suitable for agricultural activities. The 15 farms in the region are mainly used to grow rice, wheat, corn, soybeans, and other kinds of beans. Such a large-scale agricultural utilization has seriously affected the local groundwater resources.

The fluctuation variance shown in Table 5.4 is calculated using the annual rainfall data collected by all groundwater monitoring stations of Jiansanjiang Sub-bureau from 1997 to 2009, and Figure 5.11 is drawn based on that. The annual precipitation change of groundwater monitoring stations in Jiansanjiang Sub-bureau from 1997 to 2009 is less than the change of paddy field area. Even though rainfall changes have an impact on the complexity of groundwater depth, this is not the most important factor. The annual precipitation fluctuation in Nongjiang Farm 8 and Qinglongshan Farm 17 is the largest, which has the greatest influence on the change of groundwater depth. This result is consistent with that shown in the figure. The North

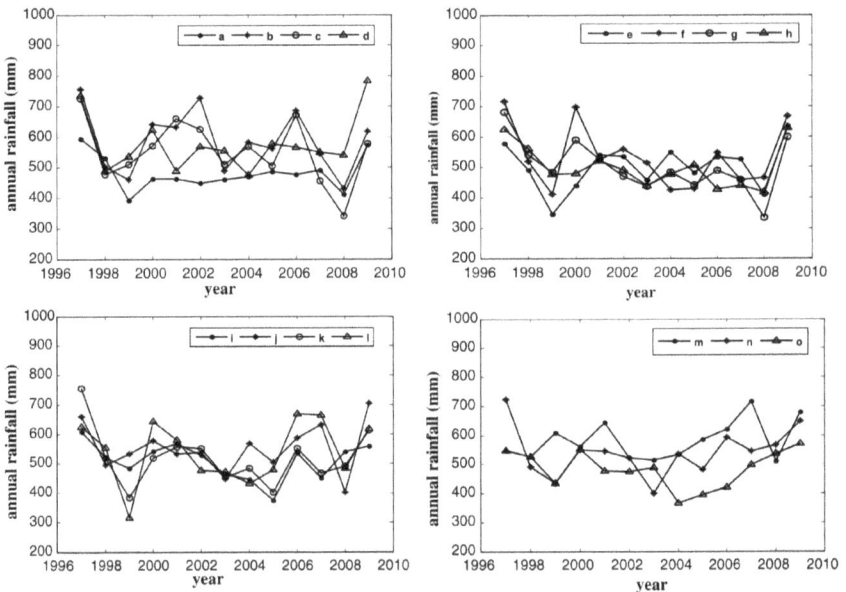

Figure 5.11. Annual rainfall of monthly groundwater depth series of each subarea in Jiansanjiang Sub-Bureau (1997–2009). (a) Qixing Farm 69; (b) Bawujiu Farm1; (c) Shengli Farm 31; (d) Qindeli Farm 7; (e) Daxing Farm 11; (f) Qinglongshan Farm 17; (g) Qianjin Farm 4; (h) Chuangye Farm 2; (i) Hongwei Farm 21; (j) Qianshao Farm 22; (k) Ministry of Qianfeng Farm; (l) Honghe Farm 6; (m) Yalvhe Farm 5; (n) Erdaohe Farm 5; (o) Nongjiang Farm 8 (Yu *et al.*, 2013).

District has the highest degree of complexity, while the South District of Jiansanjiang Sub-bureau has the smallest degree of annual precipitation fluctuation. The above analysis shows that the most important factor affecting the dynamic change of local groundwater depth is human activities, especially through changes in paddy field areas, followed by local climate change.

5.1.4 *Dynamic characteristics of groundwater system in the lower Yellow River based on nonlinear system theory*

This study was carried out by Chen and Huo (2008). The Yellow River enters the North China Plain from Taohuayu in Zhengzhou, forming a suspended river on the ground and no surface watershed. The Yellow River water is higher than the surface and plays an important role in recharging the groundwater. Its recharging influence width is 10–20 km, and the circulation depth is about 300 m. The groundwater system mainly consists of geomorphology such as washland, flood fan, low-lying land, flood plain. The loose sediments are mainly alluvial–diluvial and lacustrine. The groundwater runoff conditions are poor, and the water cycle is dominated by vertical movement. According to the buried conditions of aquifers, the groundwater system can be divided into shallow groundwater subsystem and deep groundwater subsystem. The system is rich in groundwater resources and has built large water source projects, which are important water supply sources for Zhengzhou (Zhao *et al.*, 2004).

5.1.4.1 *Groundwater system in the affected zone of the lower Yellow River*

According to the buried conditions, hydrodynamic characteristics, and exploitation of aquifers, the groundwater system in the affected zone of the lower Yellow River can be divided into shallow groundwater subsystem and deep groundwater subsystem. Two aquifers are relatively independent and have certain hydraulic connections, with little difference in water levels. Between the shallow aquifer and the deep aquifer is a relatively stable weak permeable layer composed of clay, with a depth of 20–30 m. Groundwater is mainly recharged by precipitation infiltration, Yellow River lateral infiltration, and irrigation infiltration, and discharged by evaporation and exploitation (Zhao *et al.*, 2004). The bottom of the shallow groundwater aquifer is concave from west to east, with a buried depth of 45–100 m, and the deepest part in Kaifeng County is 130 m. It is a typical "binary structure" aquifer that the upper part of it is consist of weak permeable silt interbedded silty clay, and silty sand layer, the lower part is consist of medium sand, medium-coarse sand, medium-fine sand and fine sand layer. The deep groundwater aquifer is mainly composed of medium-fine sand, fine sand, and silt. Due to different genesis, the thickness of the sand layer varies greatly, ranging from 40 to 110 m. The buried depth of the aquifer floor is 300–450 m.

Due to the intrinsic complexity in nonlinear systems, it is very difficult to construct a complete system model. Generally only the time series of one or several components of the system can be measured. At this time, the phase space of the system can be reconstructed by phase space reconstruction methods. Phase space reconstruction theory holds that the evolution of any variable in the system is determined by other components interacting with it. Therefore, the information on these related components is hidden in the development process of any component. Therefore, only one component is considered, and the observed values at some fixed delay time are treated as new dimensions so that a phase space equivalent to the original system can be constructed by the "embedding" method, and the original dynamic system can be restored in this phase space, and the properties of its attractors can be studied. It has been proved that when the embedding dimension m and the delay time τ are properly selected, the reconstructed phase space can have the same geometric and information properties as the actual dynamic system and have all the characteristics of the real space.

The dynamic characteristics of nonlinear complex systems can be discussed from the fractal characteristics, the complexity degree, and the types of system evolution. Based on a large number of empirical studies, Hurst proposed the Hurst index and R/S analysis method, which can effectively obtain whether the time series of system variables have a fractal structure and a related persistence or not (Xu *et al.*, 2002). The Hurst index of the time series is between 0 and 1, with an interval of 0.5. Time series show different characteristics in different intervals. It is a fractal Brownian motion when $H \in (0, 0.5)$, the future data of the time series tends to return to the historical point. It is standard Brownian motion when $H = 0.5$, at this time, it is characterized by a Markov chain. It is a long-term persistence and periodic cycle when $H \in (0.5, 1)$, and the time series is chaotic. Therefore, records within a certain range will last for a long period, and it is difficult to predict future changed according to the past. The time series is completely predictable when $H = 1$. At this time, the time series is a straight line, and the future can be predicted.

Fractal dimension is the main index to describe the complexity of nonlinear systems, and it can help to determine the main state variables affected the system. Therefore, the fractal dimension can be used as an index to predict the future changes of the system to a certain extent. The methods of calculating the fractal dimension include the topological dimension, Hausdorff dimension, information dimension, and correlation dimension, of which correlation dimension has the characteristics of conservatism, simplicity, and stability.

For a nonlinear system, it evolves in the phase space for a long time, and the final performance is that the trajectory either approaches a point or a closed curve, or attracted to a region in the phase space and shows an irregular motion within a certain range. The combination of Kolmogorov entropy (K) and Lyapunov exponent ($\lambda 1$) can well distinguish the evolution type of the system. The property of the system motion can be judged using K value, $K = 0$, indicating that the system moves regularly. K is equal to positive infinity, indicating that the system moves randomly. $0 < K < +\infty$, indicating that the system is in chaotic motion. If $\lambda 1 < 0$, it indicates that there are attractors in the system. $\lambda 1 = 0$ indicates that the system has limit cycles, and $\lambda 1 > 0$, it indicates that the system has strange attractors.

5.1.4.2 *Dynamic characteristics of the groundwater system influenced by the lower Yellow River*

Three long-term observation wells for shallow groundwater in the affected zone of the lower Yellow River were selected, while two long-term observation wells were selected for deep groundwater. The data series are all five-day observation data from 1981 to 2002.

5.1.4.2.1 Calculation of system characteristic index

The C-C method is used to calculate the time delay. The correlation dimension is calculated using the correlation integration method. Table 5.5 shows the calculation results of dynamic characteristic indexes of each subsystem.

5.1.4.2.2 System dynamics characteristics analysis

The fractal features of the system: In Table 5.5, the Hurst index values of each subsystem reflect the following characteristics:

(1) The Hurst index values of the two subsystems are between 0.5 and 1.0, which indicates that the time series of representative variables of the two subsystems are chaotic, that is, the two subsystems are chaotic, and the groundwater system in the influence zone of the lower Yellow River is chaotic.
(2) The Hurst index value of the shallow groundwater subsystem is smaller than that of the deep groundwater subsystem. From the shallow groundwater circulation to deep groundwater, the Hurst index value increases with the groundwater buried depth.
(3) The degree of opening and the external disturbance of the shallow groundwater subsystem are higher than that of the deep groundwater subsystem, and the Hurst index of the subsystem increases with the decrease of the degree of opening and the external disturbance, which indicates that Hurst index can describe the degree of opening and external disturbance of the system.

The complexity of the system: The fractal dimension of the shallow groundwater subsystem is about 2.52, indicating that the subsystem

Table 5.5. Chaotic characteristic index of groundwater level time series (Chen and Huo, 2008).

Subsystem	Time delay	Fractal dimension D	Embedding dimension m	Hurst Index H	Kolmogorov entropy K	Max Lyapunov index λ1
Shallow groundwater	7	2.50–2.53	7	0.91–0.92	0.054–0.058	0.0481–0.085
Deep groundwater	8	3.35 (3.36)	8	0.96 (0.97)	0.073 (0.081)	0.0584 (0.0591)

should be described by at least three main state variables. The fractal dimension of the deep groundwater subsystem is about 3.35, indicating that the subsystem should be described by at least three main state variables. Compared with the shallow groundwater subsystem, the deep groundwater subsystem is a relatively closed subsystem with less external interference. However, the correlation dimension of the time series of deep groundwater level calculated this time is larger than that of the shallow groundwater subsystem, which seems to be somewhat abnormal. It may be that the chaotic characteristics of the deep groundwater subsystem are not obvious. The Hurst index of the deep groundwater subsystem is close to 1, also confirm it.

The evolution of the system: The Kolmogorov entropy of representative variable for the groundwater system time series are shown in Table 5.5. The Kolmogorov entropy of shallow groundwater subsystem and deep groundwater subsystem yields positive numbers that, indicating that both subsystems have chaotic characteristics. The calculation results of Kolmogorov entropy, Hurst exponent, and Lyapunov exponent all show that the groundwater system has obvious fractal structure and chaotic characteristics, is a nonlinear system, and the system has the largest Lyapunov exponent that is greater than zero, which means that there is a strange attractor in the evolution of groundwater system.

5.1.5 *Complexity measurement of groundwater system in urbanization region based on permutation entropy*

This study was carried out by Zhang and Liu (2017). Harbin is the capital of Heilongjiang Province. It is the political, economic, and cultural center in the northeast China. The total area of the city is approximately $53840\,km^2$. In 2014, the total registered population was 9.94 million. Harbin has a mid-temperate continental monsoon climate, with long winter and short summer, the annual average precipitation is 569.1 mm and mainly concentrated in June to September (Li, 2000). Under the influence of underlying surface conditions (such as topography and geomorphology), the groundwater distribution in this area is uneven and the complexity characteristics are obvious.

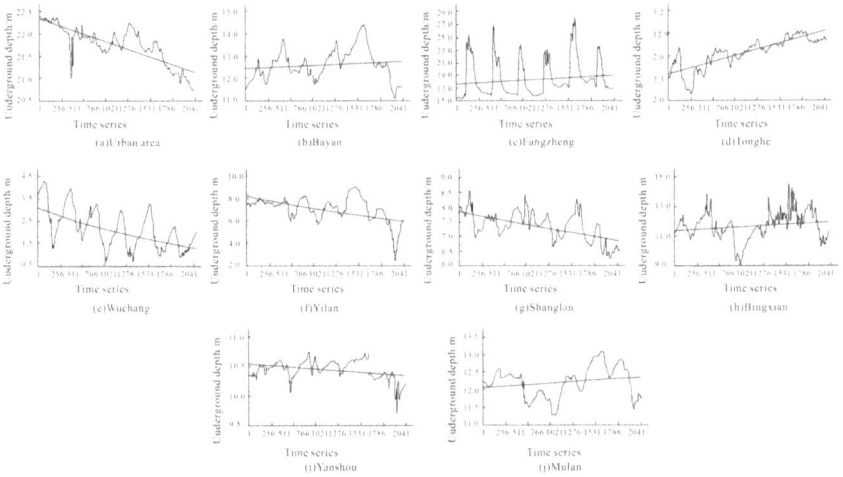

Figure 5.12.   Daily groundwater depth change in monitoring sites of Harbin City
(2008–2013) (Zhang and Liu, 2017).

Therefore, the complexity characteristics of the regional groundwater depth sequence are revealed, to establish a foundation for the prediction and rational usage of regional groundwater resources.

5.1.5.1   *Data source*

The daily groundwater depth monitoring data ($n = 2192$) from 2008 to 2013 in 10 sites were collected from Heilongjiang Meteorological Bureau. The groundwater depth of monitoring sites in Harbin shows periodic changes, shown in Figure 5.12.

5.1.5.2   *Research method: Permutation entropy*

Bandt and Pompe (2002) proposed an algorithm PE (Permutation Entropy) to measure the complexity of one-dimensional time series, which has the characteristics of simple calculation and strong anti-noise interference ability.

The specific algorithm of PE is as follows (Srinu and Mishra, 2016):

(1) Let the time series X_i, $i = 1, 2, \ldots, n$, the phase space is reconstructed and the matrix $\mathbf{X}_K$ is obtained;

$$
\begin{bmatrix}
x_1 & x_{(1+r)} & \cdots & x_{[1+(m-1)\tau]} \\
\vdots & \vdots & \vdots & \vdots \\
x_j & x_{j+\tau} & \cdots & x_{[j+(m-1)\tau]} \\
\vdots & \vdots & \vdots & \vdots \\
x_K & x_{(K+\tau)} & \cdots & x_{[K+(m-1)\tau]}
\end{bmatrix}
\qquad j = 1, 2, \ldots, K \quad (5.8)
$$

where m and τ is the embedded number of digits and delay time, respectively, while $K = n - (m-1)\tau$.

(2) Rearrange the J refactoring component $[x_j, x_{j+}, \ldots, x_{j+(m-1)\tau}]$, in X_i refactoring matrix in ascending order, to get $j_1, j_2, \ldots, j_m$, i.e.,

$$
x_{i+[j(1)-1]\tau} \leq (x_{i+[j(2)-1]\tau} \leq \cdots \leq) x_{i+[j(m)-1]\tau} \qquad (5.9)
$$

(3) Time series X_i arrangement has m! types in total. The probability of occurrence of each sequence is calculated as $P_1, P_2, \ldots, P_m$, so PE is:

$$
PE_p(m) = -\sum_{j=1}^{m} p_j ln P_j \qquad (5.10)
$$

$PE_p(m)$ represents the degree of randomness of time series X_i, and the greater the PE value, the more complex the time series.

5.1.5.3 *Results*

Using the above method, where the embedded digit m takes 4, the delay time τ takes 2, the PE values of daily groundwater buried depth sequence in each Harbin monitoring site are computed using the MatLabR 2010B software program, the result of which is shown in Table 5.6.

If the PE value of the groundwater depth sequence is larger, the complexity is stronger and the predictability of the groundwater depth sequence is reduced. From Table 5.6, it can be seen that the predictability of groundwater depth sequence in Harbin is

Table 5.6. Entropy of daily groundwater buried depth sequence in Harbin City based on PE (Zhang and Liu, 2017).

Stations	Entropy value	Ranking	Stations	Entropy value	Ranking
Urban District	2.2524	7	Bayan	2.7605	10
Binxian	2.6503	9	Fangzheng	1.9062	3
Mulan	1.7446	1	Shangzhi	1.9496	5
Tonghe	1.9126	4	Wuchang	1.8936	2
Yanshou	1.9550	6	Yilan	2.5585	8

Table 5.7. Classification of complexity for daily groundwater buried depth sequence (Zhang and Liu, 2017).

Area	Entropy value	Range of values	Grade
Urban District	2.2524	2.400–1.900	II
Bayan	2.7605	2.900–2.400	I
Binxian	2.6503	2.900–2.400	I
Fangzheng	1.9062	2.400–1.900	II
Mulan	1.7446	1.900–1.400	III
Shangzhi	1.9496	2.400–1.900	II
Tonghe	1.9126	2.400–1.900	II
Wuchang	1.8936	1.900–1.400	III
Yanshou	1.9550	2.400–1.900	II
Yilan	2.5585	2.900–2.400	I

Mulan > Wuchang > Fangzheng > Tonghe > Shangzhi > Yanshou > Urban District > Yilan > Binxian > Bayan. According to PE value, complexity is divided into three levels. The entropy value for Grade I is between 2.900 and 2.400, that for Grade II is between 2.400 and 1.900, and for Grade III is between 1.900 and 1.400, as shown in Table 5.7, and the complex spatial distribution of groundwater depth sequence is drawn, as shown in Figure 5.13.

From Figure 5.13, it can be seen that the complexity of groundwater depth sequence in Bayan, Binxian, and Yilan regions is Grade I, which indicates that it is difficult to predict groundwater resources in these three regions and there are many influencing factors. The complexity of groundwater depth sequence in the five regions of the urban area, Fangzheng, Shangzhi, Tonghe, and Yanshou is Grade II, which indicates that the prediction of groundwater resources in these

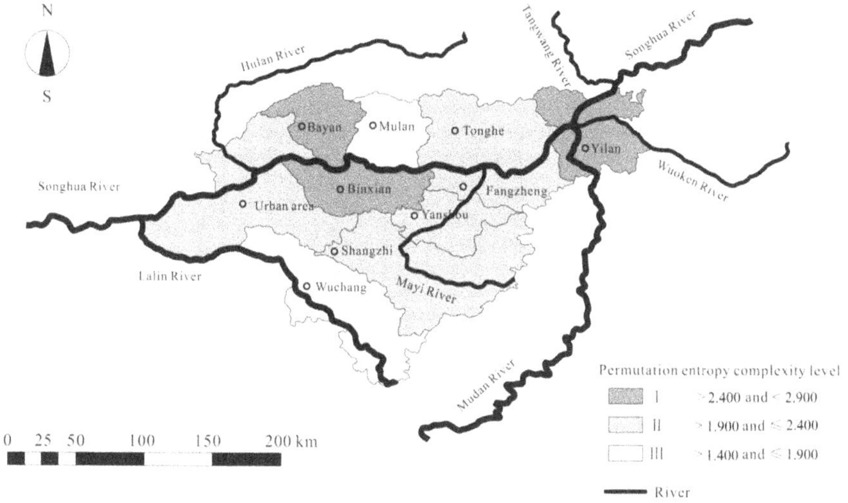

Figure 5.13. Spatial distribution of complexity of daily groundwater depth sequence in Harbin (Zhang and Liu, 2017).

five regions is generally difficult and the number of influencing factors is general. The complexity of groundwater depth sequence in Mulan and Wuchang regions is Grade III, which indicates that groundwater resources in these two regions are easy to predict and have few influencing factors.

5.1.5.4 *Discussion*

In this paper, two underlying surface conditions (mountain area and water area) in the Harbin are selected to analyze the correlation between the two conditions and the complexity of groundwater depth sequence and to explore their influence on the complexity of groundwater hydrological system. The proportion of mountain area and water area are calculated, and divided into three levels, as shown in Table 5.8. The mountain area is Grade I, Grade II, and Grade III when the area proportion is in the range of 60–80%, 35–60%, and 0–35%, respectively, while the water area is Grade I, Grade II, and Grade III when the area proportion is in the range of 4.9–67.0%, 2.5–4.9%, and 1.0–2.5%, respectively. The spatial distribution of mountain area proportion and water area proportion are drawn on

Table 5.8.   Area proportion and classification of mountainous and water in Harbin (Zhang and Liu, 2017).

Area	The ratio of mountainous and hilly areas (%)	Grade	The proportion of water area (%)	Grade
Mulan	46	II	2.1	III
Shangzhi	80	I	5.0	I
Bayan	33	III	1.8	III
Binxian	37	II	3.3	II
Yanshou	70	I	5.1	I
Tonghe	78	I	3.0	II
Wuchang	43	II	1.4	III
Fangzheng	80	I	2.8	II
Yilan	36	II	2.6	II
Urban District	0	III	2.9	II

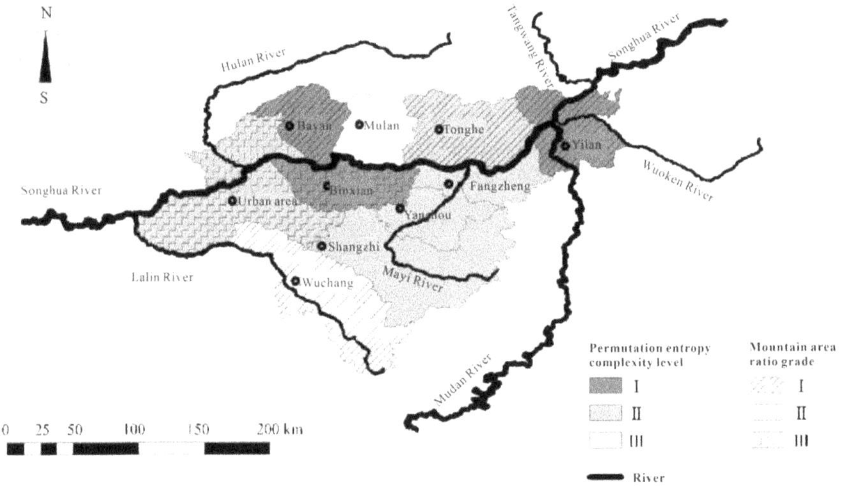

Figure 5.14.   Spatial distribution of mountain area in Harbin (Zhang and Liu, 2017).

the complex spatial distribution base map of daily groundwater depth sequence in Harbin region, as shown in Figures 5.14 and 5.15.

From Figure 5.14, it can be seen that the complexity grade of groundwater depth sequence of Bayan, Yilan, Shangzhi, Yanshou, and Fangzheng is consistent with the proportion grade of the mountain area, which indicates that the mountain topography in these five

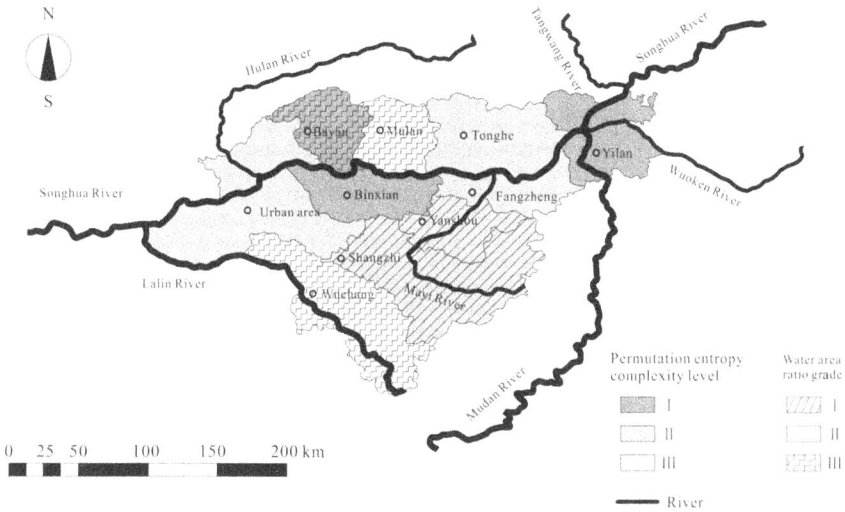

Figure 5.15. Spatial distribution of water area proportion in Harbin (Zhang and Liu, 2017).

regions has a great influence on the groundwater hydrological system. The complexity grade of groundwater buried depth sequence in Binxian and Wuchang is two grades different from that of mountain area proportion, which indicates that the mountain topography in these two regions has little influence on the groundwater hydrological system. The difference between the two indicators in the other regions is one grade, which indicates that the mountain topography in these three regions has a general influence on the groundwater hydrological system.

From Figure 5.15, the groundwater buried depth sequence complexity level of Bayan and Yilan is consistent with the water area grade, indicating that the water area in these two regions have a greater impact on the groundwater hydrological system; Groundwater depth sequence complexity levels of Binxian and Wuchang are 2 grades different from that of water area. The water area of the two regions have a smaller impact on the groundwater system. For the rest of regions, the difference of two indicators is only one grade, indicating that water area in these six regions have a general impact capacity on the groundwater hydrological systems.

To sum up, the mountain topography and water area in the Harbin have an important influence on the complexity of groundwater depth sequence, but there are differences, and the influence of mountain topography is greater than that of water area. Therefore, the underlying surface conditions have a certain influence on the complexity of the groundwater hydrological system.

5.2   Case Study Based on Model and Numerical Simulation

5.2.1   *Multi-dimensional dynamic system model of groundwater resources in karst area, taking Zaozhuang City as an example*

This study was carried out by Han *et al.* (2005). The multi-dimensional dynamic system of groundwater resources consists of input, output, and hydrogeological entities. The real multi-dimensional dynamic system of groundwater resources is a very large and complicated system. There are many influencing factors, and the observation and research of groundwater are far less mature and in depth than that of surface water. Therefore, the characteristics of a multi-dimensional dynamic system of groundwater resources are generally described by models to describe the characteristics of entity systems and their dynamic changes. At present, the commonly used models are all abstract and generalized mathematical models, that is, the hydrogeological entity prototype is summarized by mathematical language, which is the abstraction of the essential attributes of the prototype. This multi-dimensional dynamic system model of groundwater is not only an optimal management model but also a prediction model of groundwater dynamics. As an optimal management model, it can provide the optimal groundwater exploitation required to meet various constraints under different recharge and groundwater level. As a prediction model, it can also predict the dynamic changes of groundwater level under different recharge and exploitation conditions, to guide the operation, management, and regulation of the system and achieve the purpose of optimal development and utilization of groundwater resources.

In this case, the system model is established using centralized parameters. The aquifer is regarded as a complete hydrogeological entity. The system parameters are expressed in a centralized form. Only the time history changes of the parameters are considered. The model is established and the parameters are determined according to the measured time series of input and output. Thus, it can be used for system management and dynamic prediction.

5.2.1.1 *Multi-dimensional dynamic system model for groundwater resources*

The time series analysis method is applied to simulate the groundwater resource system, and it is based on the water balance equation, but it also makes the traditional water balance method have dynamic properties. Therefore, it can be used to simulate the dynamic characteristics of groundwater resources system, especially when the data is insufficient for the analytical method or numerical method, it is especially effective and suitable. It is economical and practical, and easy to popularize and apply.

5.2.1.1.1 Model structure setting

Here, the multi-dimensional dynamic system of groundwater resources is regarded as an underground reservoir, and its water balance problem is studied. The form of its water balance equation is:

$$\Delta H = b_0 + b_1 P + b_2 M + b_3 Q_1 - b_4 Q_2 + b_5 W_s$$
$$+ b_6 W_g - b_7 E - b_8 Q_k \tag{5.11}$$

In the formula, P is the precipitation (mm) in the calculation period; M is the channel penetration recharge depth (negative when groundwater recharges the river) (m); W_s is the surface water consumption in the calculation period, represented as water depth (m); W_g is the amount of groundwater mining (m); E is phreatic water evaporation (m); Q_1 and Q_2 are groundwater lateral inflow and outflow (m^3); Q_k is base flow (m^3); $b_0, b_1, b_2 \cdots b_7, b_8$ are model parameters.

5.2.1.1.2   Parameter Estimation

Equation (5.11) is for a certain period. After the calculation period is selected, the measured time series data of the selected independent variables and dependent variables are taken into Equation (5.11), thus establishing a time series equation group. To facilitate the expression, the independent variables in the equations are represented by $x(i, j)$, while the dependent variables are represented by $y(j)$, where i represents the serial number of the independent variables and j represents the time serial number. Therefore, the time series equations can be expressed as:

$$y(1) = b_0 + b_1 x(1, 1) + b_2 x(2, 1) + b_3 x(3, 1) - b_4 x(4, 1) \qquad (5.12)$$
$$+ b_5 x(5, 1) - b_6 x(6, 1) - b_7 x(7, 1) - b_8 x(8, 1)$$

$$y(2) = b_0 + b_1 x(1, 2) + b_2 x(2, 2) + b_3 x(3, 2) - b_4 x(4, 2) \qquad (5.13)$$
$$+ b_5 x(5, 2) - b_6 x(6, 2) - b_7 x(7, 2) - b_8 x(8, 2)$$

$$\cdots$$

$$y(n) = b_0 + b_1 x(1, n) + b_2 x(2, n) + b_3 x(3, n) - b_4 x(4, n) \qquad (5.14)$$
$$+ b_5 x(5, n) - b_6 x(6, n) - b_7 x(7, n) - b_8 x(8, n)$$

where n is the sequence length. The least-squares method is used to solve the equations, thus the model parameter b can be estimated.

5.2.1.1.3   Model checking

After the system model is established, the validity of the model needs to be checked before use. Usually, the following two aspects need to be checked:

(1) *Significance test of a linear relationship*: The structure of the system mathematical model is a multiple linear regression equation, so firstly it should be tested for significance to determine the degree of linear correlation between dependent variables and independent variables to confirm the practicability of the multiple linear regression equation.
(2) *The test of fit with the measured data*: The model should well simulate the dynamic change of groundwater. Therefore, the measured values of independent variables are input into the model, then the outputting data of dependent variable are compared

with the measured data to test fitting goodness, that is simulation accuracy. After the model has passed the above two tests and is confirmed to be valid, the model can be used for the prediction, regulation, and management of multi-dimensional dynamic systems of groundwater resources.

5.2.1.2 *Model of a multi-dimensional dynamic system of groundwater resources in Zaozhuang karst region*

5.2.1.2.1 Setting of the model structure of Zaozhuang karst zone system

This test area is located in the south of Zaozhuang city, with a total area of $171.8\,\mathrm{km}^2$. The modeling series takes years as the calculation period and takes the 20-year synchronous time series data from 1981 to 2000 to establish the model.

It is considered that the multi-dimensional dynamic system of groundwater resources studied is quasi-stochastic, that is, the physical characteristics of the system do not change with time. In the process of exploitation, groundwater recharge and consumption have generally formed a dynamic equilibrium state, so the difference between inflow and outflow of groundwater can be regarded as a constant. Given the large buried depth of groundwater because of mining, phreatic evaporation is ignored in the calculation. At the same time, the balance item Ws will not be introduced in this area without considering the transfer of water from outside the country in a short time. Therefore, the main factors affecting the change of groundwater level in the system are precipitation P, river infiltration M and groundwater exploitation W_g. Therefore, Equation (5.11) can be simplified to:

$$\Delta H = b_0 + b_1 P + b_2 M - b_3 W_g \qquad (5.15)$$

In this formula, b_0, b_1, b_2, and b_3 are model parameters.

To reduce the dimension and simplify the convenience of calculation, Equation (5.15) can be simplified as:

$$\Delta H = b_0 + b_1 P - b_2 W_g \qquad (5.16)$$

5.2.1.2.2   Estimation of system model parameters

According to the set model (5.16), it is necessary to calculate the synchronous time series data of groundwater level variation ΔH, precipitation P, and groundwater exploitation Wg in the system during the period.

The variation amplitude ΔH of groundwater level takes the difference between the groundwater level on December 31 of the current year and the groundwater level on December 31 of the previous year for six representative groundwater dynamic observation wells in the test area. Precipitation P is the average annual precipitation of the four precipitation stations in the test area. Groundwater exploitation W_g takes the total exploitation of industrial, agricultural, human, and animal water supply in the test area. The equations can be solved using the least-squares method, so the final system model is as follows:

$$\Delta H = -9.54 + 0.01149\,P - 0.000171\,W_g \qquad (5.17)$$

5.2.1.2.3   Inspection of system model

The significance test of the above linear regression model (5.17) can be tested by correlation coefficient test, F test, and so on. Taking sample complex correlation coefficient γ_0 as a statistic, intermediate value of the correlation coefficient during determining the model parameters is used to get $\gamma_0 = 0.76$.

The model of this system is a binary regression equation with a total number of variables of 3, sample capacity (i.e., the length of the series) $n = 20$, so the degree of freedom is $n - 3 = 17$. At confidence $\alpha = 0.05$, that is, 95% confidence level, the critical value of the correlation coefficient $\gamma_\alpha = 0.482$, apparently $\gamma_0 > \gamma_\alpha$, it is considered that the regression equation of this model is significant and that the linearity is strong.

In addition to the significance test of linear relationship for the system model, it is necessary to further test the goodness of fit for the system model. Inputting the calculated data P and Wg (5.17), calculate the output value ΔH and then compare it with the measured value ΔH to test the goodness of fit of the system model. After comparison and test, the fitting is satisfactory, and the model set for this test area can express the ΔH–P–W_g relationship well. The system model is effective and reliable and has a practical value. It can directly be used for the prediction, regulation, and management of a multi-dimensional dynamic system of groundwater resources in the Zaozhuang karst area.

5.2.2 *Regional groundwater system complexity measurement and its impact on model prediction accuracy*

This study was carried out by Yu and Liu (2012). Groundwater resources, as an important part of water resource circulation, are closely related to human production and life, and their fluctuations directly affect the daily life of residents and even the development of the national economy. However, due to the influence of the natural environment and human factors, its activities show many complex characteristics such as nonlinearity and dynamics, irreversibility, self-adaptability, accumulation effect (initial value sensitivity), strange attraction, and structural self-similarity (Wang *et al.*, 2005), which makes groundwater level declining and water quality deterioration more and more frequent. However, previous scholars often ignore the influence of the internal complexity of the system in the study of groundwater resources fluctuation rules, resulting in the analysis results often deviating from the actual situation. Under this background, it is necessary to study the evolution rules of the regional groundwater system complexity to provide a scientific basis for grasping the dynamic rules of the groundwater system and realizing the rational utilization of groundwater.

There are two key points in the study of system complexity, that is, starting from system external driving mechanism or internal constituent elements, the research on a system external driving mechanism often leads to deviation of analysis results due to numerous driving factors and difficulty in grasping the actual situations. In contrast, it is more reasonable and reliable to analyze system complexity by using internal changes in system constituent elements.

There are many methods to measure system complexity, among which Kolmogorov entropy, as a characteristic quantity to quantitatively describe the degree of chaos in the system, is a simple and effective method to measure system complexity. With the development of chaos theory in recent years, this method has been less applied in the complexity diagnosis of the groundwater system. The complexity of the groundwater system and the groundwater prediction model are important to guide the rational utilization of groundwater resources. This has widely concerned hydrological workers for a long time, but the research on the correlation between groundwater complexity and Kolmogorov entropy is a blank. Studying the internal relationship between them is of great significance for the improvement and screening of prediction models.

5.2.2.1   *Data collection*

Jiansanjiang Sub-bureau is located in the hinterland of Sanjiang Plain and has 15 farms under its jurisdiction, as shown in Figure 5.16. The area is rich in water resources and has a mild and humid climate, flat terrain, and fertile soil. It is an important commodity grain and green organic food base in China. In recent years, as a result of the excessive pursuit of economic benefits, the paddy field area has increased rapidly, and the use of irrigation water (dominated by groundwater) has increased. At the same time, the necessary water control works is lacking, both have led to a general decline in groundwater level in the study area, overexploitation of groundwater happened occasionally. Combined with constraints such as the

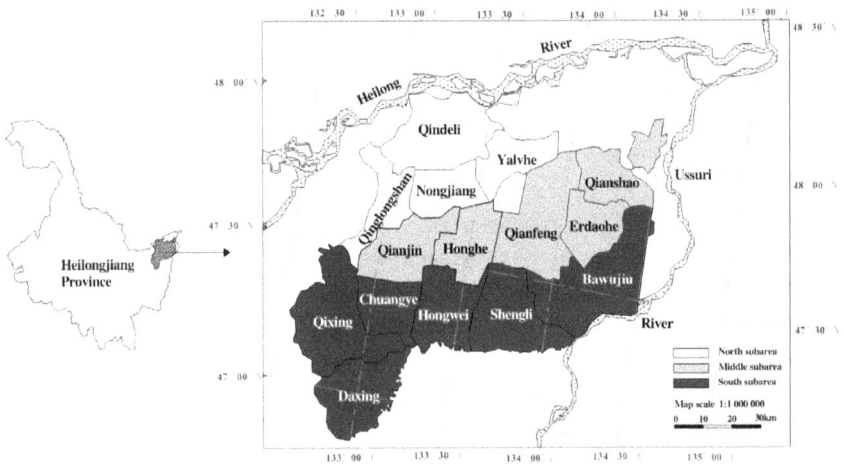

Figure 5.16.   Location map of Jiansanjiang Sub-bureau of Heilongjiang state farms (Yu and Dong, 2012).

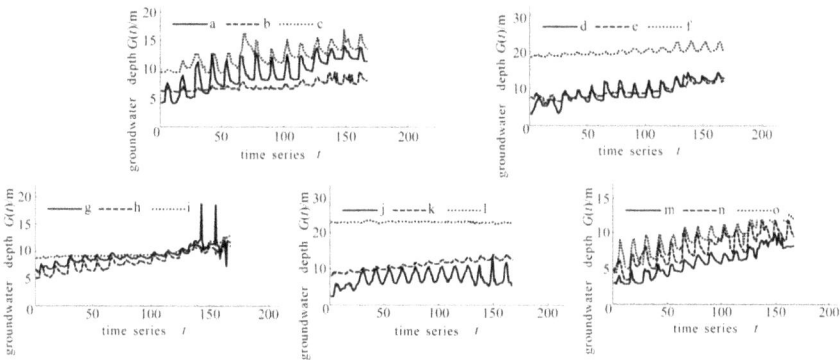

Figure 5.17. Monthly change of groundwater depth in Jiansanjiang Sub-bureau farms (1997–2010) (Yu and Dong, 2012). (a) Chuangye Farm 2; (b) Yalvhe Farm 5; (c) Hongwei Farm 1; (d) Nongjiang Farm 8; (e) Qinglongshan Farm 15; (f) Qindeli Farm 7; (g) Honghe Farm 6; (h) Qixing Farm 69; (i) Qianjin Farm Team 4; (j) Bawujiu Farm 1; (k) Qianshao Farm 22; (l) Shengli Farm 37; (m) Daxing Farm 11; (n) Erdaohe Farm 3; (o) Ministry of Qianfeng Farm.

natural environment and geological conditions, has led to the increasing complexity of groundwater systems in the region. Therefore, it is necessary to research the evolution of the complexity of groundwater systems in the Jiansanjiang Sub-bureau to provide a scientific basis for the rational exploitation and use of water resources in the region.

Based on the monthly observation data of groundwater depth of 15 farms in Jiansanjiang Sub-bureau from 1997 to 2010 ($n = 168$), the fluctuation and change of groundwater depth sequence with time is shown in Figure 5.17. As can be seen from Figure 5.17, the monthly groundwater depth curves of each farm show non-stationary random change characteristics to different degrees.

5.2.2.2 Research methods

5.2.2.2.1 Phase space reconstruction

The phase space reconstruction theory proposed by Packard *et al.* (1980), that is introducing chaos theory into nonlinear time series analysis. The central idea of this theory is to reconstruct the system by transforming the one-dimensional time series into multi-dimensional phase spaces with the same topology through introducing delay time τ and embedding dimensions m. Let the observation

Table 5.9. Delay time τ calculation results. (Yu and Dong, 2012).

Farm	Chuangye	Yalvhe	Hongwei	Nongjiang	Qinglongshan
	40	52	51	41	56
	Qindeli	Qianfeng	Honghe	Erdaohe	Daxing
	54	55	62	53	4
	Shengli	Qianshao	Bawujiu	Qixing	Qianjin
	63	26	63	53	53

Delay time τ is indicated along the left side of the table.

time series be $x_i = x(t_i)$, $t_i = t_0 + i\Delta t$, $i = 1, 2, \ldots, N$, select delay time τ, to structure m dimensional phase space. Its definition is:

$$y_i = \{x_i, x_{i+\tau}, x_{i+2\tau}, \ldots, x_{i+(m-1)\tau}\} \tag{5.18}$$

$$[i = 1, 2, \ldots, N - (m-1)\tau] \tag{5.19}$$

The selection of the delay time τ and embedding dimension m is of great significance for phase space reconstruction. The calculation for delay time τ mainly includes the autocorrelation function method and the mutual information method. This work adopts the autocorrelation function method and the formula is as follows:

$$r_k = \frac{\sum\limits_{t=1}^{n-k} (x_i - x)(x_{i+k} - x)}{\sum\limits_{t=1}^{n} (x_t - x)^2} \tag{5.20}$$

where r_k is the k order autocorrelation coefficient, x_t is the mean value of x, when the autocorrelation coefficient approaches 0, k is the desired τ. The calculation of the delay time τ in the sequence of groundwater buried depth for farms is shown in Table 5.9.

5.2.2.2.2 The determination of correlation integral

Vectors whose distance is less than a given positive number r, is called an associated vector. The Theiler window correlated integral correction formula is used to define the function $C_{(r)}$ as:

$$C_m(r) = \frac{2}{(N+1-w)(N-w)} \sum_{n=w}^{N} \sum_{i=1}^{N-n} H(r - \|x_{i+n} - x_i\|)$$

$$= \frac{2}{(N+1-w)(N-w)}$$

$$\times \sum_{n=w}^{N} \sum_{i=1}^{N-n} H \left\{ r - \left[\sum_{k=1}^{m-1} (x(i+n+k) - x(i+k)^2 \right]^{1/2} \right\} \tag{5.21}$$

In the formula, $N = n - m + 1$; for the function $H(a)$, $H(a) = 1$ when $a > 0$, $H(a) = 0$ when $a \le 0$; r for a given positive decimal, is called critical distance; $\|\cdot\|$ is the Euclidean norm; $C_m(r)$ is the probability that distance of two points in the m dimensional phase space is less than r.

5.2.2.2.3 Kolmogorov entropy

Kolmogorov entropy gives an upper and lower limit of the average amount of information generated by the orbit in a unit time, so it can be regarded as a index to judge the complexity of the system. $0 < K < \infty$ illustrates the complexity of the sequence, and the greater K value indicates the more serious complexity of the system. The Kolmogorov entropy formula is calculated by G-P algorithm as:

$$K = \frac{1}{\tau} \ln \frac{C_m(r)^2}{C_{m+1}(r)^2} \tag{5.22}$$

$C_m(r)$ is the correlation integral when the embedding dimension is m; $C_m + 1(r)$ is the correlation integral when the embedding dimension is $m + 1$.

Typically, it is taking stable value that K varies with $(m + 1)$ as the estimated K. Figure 5.18 shows that the Kolmogorov entropy of groundwater buried depth sequence in Qianshao Farm 22 changes with $m + 1$ when $r = 4$. It can be seen in the figure, when $(m + 1) = 13$, that is $m = 12$, the variation is almost stable, at which point $K = 0.0027$ illustrates the complexity of the sequence.

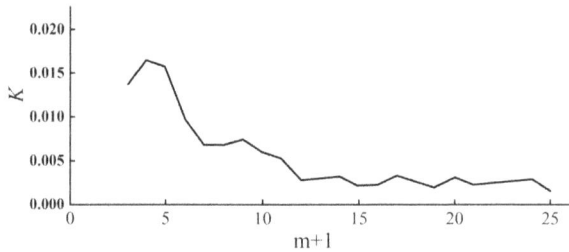

Figure 5.18. Kolmogorov entropy change curve of groundwater buried depth sequence in Qianshao Farm 22 with $(m + 1)$ (Yu and Dong, 2012).

Table 5.10. Complexity of groundwater depth sequence in each farm of Jiansanjiang Sub-bureau (Yu and Dong, 2012).

Partition	Monitoring location	Well number	K	Complexity sort	R
North District	Nongjiang Farm 8	No.8	0.1078	2	0.0316
	Qinglongshan Farm 15	1508	0.0083	3	
	Qindeli Farm 7	407001	0.0037	11	
	Yalvhe Farm 5	00-05	0.0066	4	
Central District	Qianshao Farm 22	No. 13	0.0027	13	0.0041
	Honghe Farm 6	601	0.0064	5	
	Qianjin Farm 4	4026	0.0053	7	
	Erdaohe Farm 3	2 1 17	0.0016	15	
	Qianfeng Farm	Feng-1 21	0.0042	8	
South District	Chuangye Farm 2	08-cyj	0.0055	6	0.0212
	Bawujiu Farm 1	101	0.0038	10	
	Qixing Farm 69	966901	0.1079	1	
	Daxing Farm 11	20	0.0031	12	
	Shengli Farm 37	No. 37	0.0025	14	
	Hongwei Farm 1	9.66E+09	0.0041	9	

5.2.2.3 *Complexity measurement results and analysis*

By the above method, together with data from 15 farms in the Jiansanjiang Sub-bureau from month-by-month history observations, K values for each farm were further calculated and the complexity is sorted, as shown in Table 5.10. At the same time, GIS software is used to visualize the complexity calculation results, and the spatial distribution map of the monthly groundwater depth sequence complexity of each farm is obtained, as shown in Figure 5.19.

From Table 5.10 and Figure 5.19, it can be seen that the groundwater depth sequence of each farm in Jiansanjiang Sub-bureau has different degrees of complexity. Among them, Qixing Farm has the strongest complexity, followed by Nongjiang Farm, and Erdaohe Farm has the weakest complexity, showing the zoning characteristics of high in the north, low in the middle, and middle in the south. Many factors lead to the complexity. First, the paddy field area of each farm has increased in recent years. As of 2009, the water consumption of paddy field has reached 252.709 million m^3, accounting for 98.4% of the total groundwater use, leading to a sharp increase in agricultural water use. Also, agricultural water use is mainly based on groundwater exploitation and the lack of necessary water control projects, making the groundwater level in each district continue

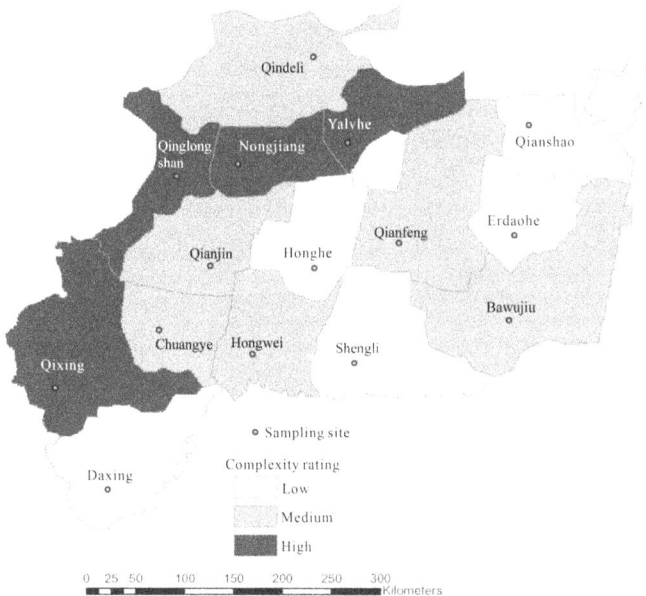

Figure 5.19. Spatial distribution map of monthly groundwater depth sequence complexity of 15 Farms in Jiansanjiang Sub-bureau.

Note: Complexity classification standard: high (0.0065–0.1080); medium (0.0035–0.0065); low (0.0015–0.0035) (Yu and Dong, 2012).

to decline, which may be the main factor leading to the complexity of the groundwater system. Secondly, due to different natural geological conditions, the distribution of precipitation is uneven and drought frequently occurs, which leads to groundwater resources cannot being effectively recharged, this may also be an important factor leading to the complexity of groundwater system. Also, the changes in residential water use and industrial water use also affect the dynamic changes of groundwater to a certain extent.

5.2.2.4 *Relevance analysis of system complexity and model predictive accuracy*

The EDM–RBF neural network model is a kind of predictive model that combines empirical mode decomposition (EDM) and radial basis function (RBF) neural network. EDM has more local representation ability than wavelet analysis, and RBF neural network has more advantages in convergence speed, extrapolation ability, and nonlinear mapping than BP neural network. The combination of the two methods can further improve the prediction accuracy of the model and has broad application prospects in groundwater depth prediction.

Combined with RBF neural network data, the EDM method is used to obtain intrinsic mode function (IMF) components and trend items of hydrological time series. RBF neural network is used to predict each IMF and trend items. The IMF component and trend items prediction values are superimposed to obtain the original sequence prediction values.

According to the above idea, the fitting value of the EMD–RBF neural network model can be obtained. Figure 5.20 shows the fitting results of groundwater depth between Qixing Farm and Chuangye Farm from 1997 to 2010. As can be seen from the curves, the fitting effects of the two are different. The specific differences can be analyzed through a model accuracy test. Table 5.11 shows the model accuracy test results of ESD–RBF neural network for groundwater depth of 15 farms in Jiansanjiang Sub-bureau.

As can be seen in Table 5.11, the EMD–RBF neural network model for each farm in Jiansanjiang Sub-bureau has met qualified standards, and the small frequency error P is 1, however the posterior difference ratio of fitting effect C, relative mean square error

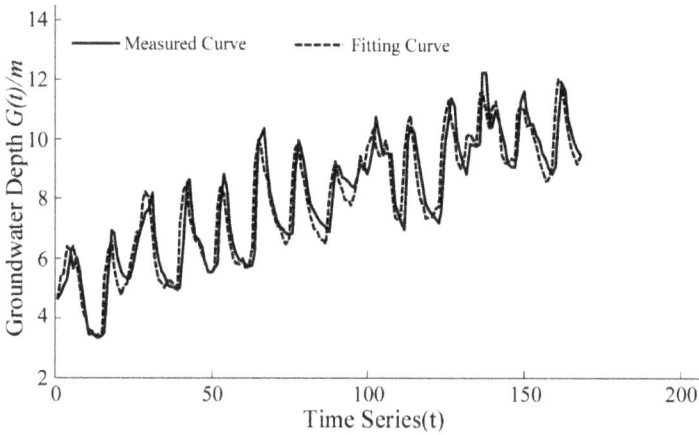

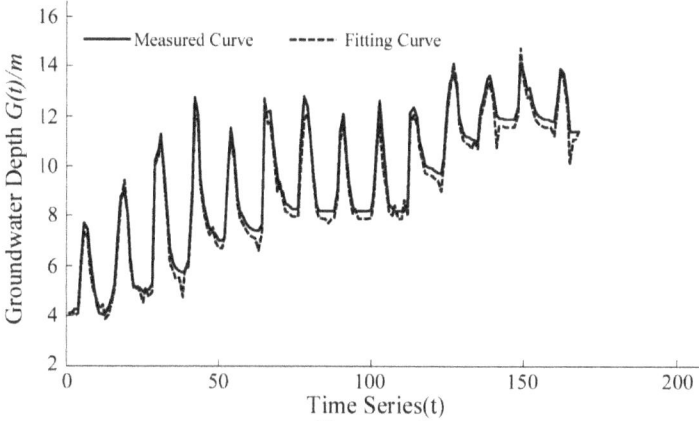

Figure 5.20.   Fitting curve of month-by-month groundwater buried depth based on EMD–RBF neural network model for (a) Qixing Farm and (b) Chuangye Farm (Yu and Dong, 2012).

E1, and fitting accuracy E2 show some differences. Thus the work explores the internal connection of complexity and model accuracy test in different farm situations. The results of each farm complexity is sorted into the sequence W1, the posterior difference ratio C, relative mean square error E1, fitting accuracy E2 results put into the sequence W2, W3, W4, W4, self-correlation analysis is carried out

Table 5.11.   EMD–RBF neural network model precision test for ground-water month-by-month buried depth in 15 farms in Jiansanjiang Sub-bureau (Yu and Dong, 2012).

Farm	Fitting effect index			
	C	P	E1 (%)	E2 (%)
Chuangye Farm	0.1934	1	7.16	69.37
Yalvhe Farm	0.2417	1	8.03	67.32
Hongwei Farm	0.1341	1	7.14	79.34
Nongjiang Farm	0.3133	1	8.43	65.80
Qinglongshan Farm	0.2648	1	8.17	67.71
Qindeli Farm	0.1060	1	6.91	85.36
Qianfeng Farm	0.1521	1	7.38	76.54
Honghe Farm	0.2145	1	7.92	68.15
Erdaohe Farm	0.0541	1	5.60	94.64
Daxing Farm	0.0913	1	6.87	87.01
Shengli Farm	0.0648	1	6.34	90.12
Qianshao Farm	0.0736	1	6.71	88.41
Bawujiu Farm	0.1253	1	7.06	80.17
Qixing Farm	0.3617	1	8.90	59.74
Qianjin Farm	0.1643	1	7.54	71.83

Note: C, posterior difference ratio; P, small frequency error; E1, relative mean square error; E2, fitting accuracy.

on W1, and its interrelated analysis with W2, W3, W4, W4, are also carried as shown in Figures 5.21 and 5.22.

According to the analysis in Tables 5.10 and 5.11, the higher the complexity, the larger the ratio of posterior test difference ratio and relative mean square error, the higher the fitting accuracy. Further analysis of autocorrelation and cross-correlation between data, can be seen in Figures 5.21 and 5.22. It shows that the cross-correlation curves and autocorrelation curves of W1, W2, W3, and W4 show random fluctuation characteristics, and the correlation coefficients all reach the maximum value when $t = 11$, with the cross-correlation coefficients of 41.81%, 37.09%, and -34.32%, respectively. This shows that W1, has a great correlation with W2, W3 and W4, and the complexity of the groundwater system has a direct impact on the prediction accuracy of the EMD–RBF neural network model. The reason may be that each driving factor of complexity is random, changeable, and unpredictable, which leads to an increase in the unpredictability

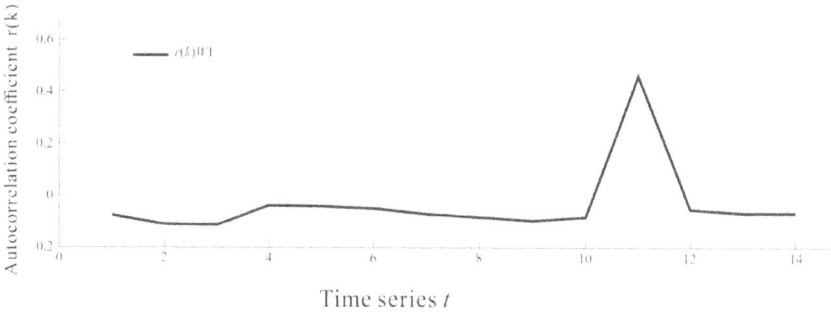

Figure 5.21. Autocorrelation diagram of month-by-month underground water deep complexity sequencing in each farm in Jiansanjiang Sub-bureau (Yu and Dong, 2012).

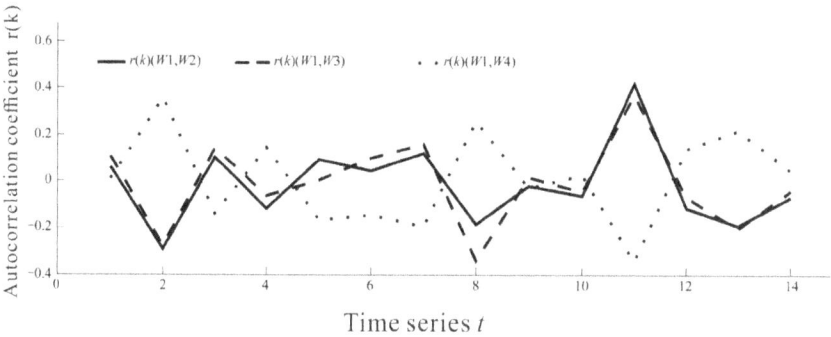

Figure 5.22. Complexity sort sequence and EMD–RBF neural network model post comparative difference (Ratio C, Relative Medium Error E1, Fit Accuracy E2 Sort Sequence) (Yu and Dong, 2012).

of the system sequence and a decrease in the prediction accuracy of the model.

Given the above problems, complexity should be taken as a variable in system analysis, the mechanism of influence of complexity on model prediction accuracy should be further studied, and methods to reduce the influence of system complexity should be explored to improve model prediction accuracy. At the same time, complexity can be taken as a screening factor for model, to obtain an optimal prediction model for a certain sequence. In a word, complexity is of great significance to the improvement of the prediction model and the prediction accuracy of groundwater variation.

5.2.3   Effects of research complexity on uncertainty in modeling of groundwater flow

This study was carried out by Samani *et al.* (2018). Uncertainty is a major problem in groundwater simulation, which may be caused by an insufficient understanding of underground systems and natural changes in underground. This incomplete understanding leads to the uncertainty, including conceptual model uncertainty, parameter uncertainty, and scenario uncertainty (Meyer *et al.*, 2007; Sun 1994). Also, the increase of model parameters (over-parameterization) will lead to the uncertainty of the model. One way to reduce uncertainty is to choose a model that compromises between a low number of parameters and a high level of performance, that is, a simple model (Hill and Tiedeman 2007). Generally speaking, the model should be as simple as possible, but it should also be noted that oversimplification will reduce the efficiency of the model (Hunt *et al.*, 2007). To evaluate the uncertainty and complexity of conceptual models, we need different conceptual models to explain the groundwater processes and the hydrogeological data (Meyer *et al.*, 2007; Rojas *et al.*, 2008; Ye *et al.*, 2004, 2008, 2010). Model averaging is a method to solve the uncertainty of the model, which has attracted attention of researchers (Dai and Ye 2015; Hojberg and Refsgaard 2005; Liu *et al.*, 2016; Neuman 2003; Poeter and Anderson 2005; Ye *et al.*, 2004). To explore the complexity of the model in the groundwater Ajabshir aquifer in eastern Azerbaijan, Iran was selected as a case study based on comprehensive hydrogeological research and modeling.

The study site is Ajabshir Plain, covering an area of $130\,\mathrm{km}^2$ as part of the catchment area of Lake Urmia in eastern Azerbaijan, Iran. The plains border the Azarshahr Plain in the north and northwest, the Maragheh-Bonab Plain in the east and south, and the Lake Urmia in the south and southwest (Figure 5.23). According to Emberger climate classification (Emberger, 1969), the study area is characterized by semi-arid and cold, with an average precipitation of about $328\,\mathrm{mm/year}$, an average temperature of about $8.7^\circ\mathrm{C}$, and an annual average relative humidity of 59%.

The study area is located in the tectonic plate collision zone, giving rise to the Alborz and Zagros Mountains and the volcanic Urmia-Dokhtar area. Geological outcrops in this area are mainly shale, sandstone, and volcanic rocks. Various sedimentary units in

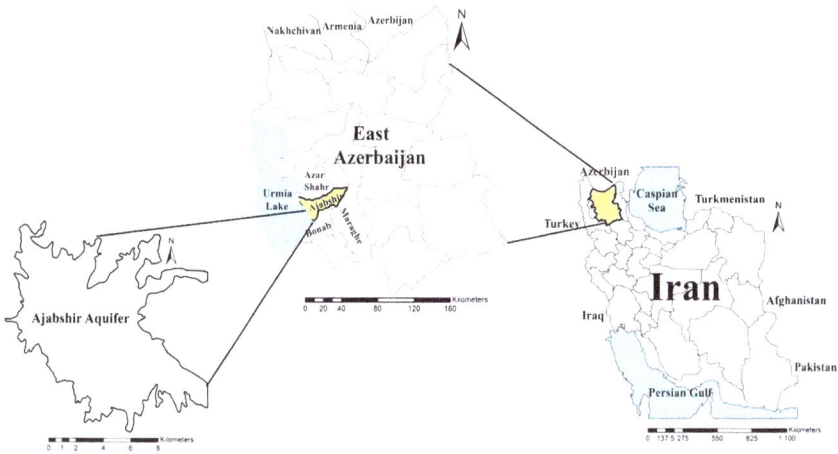

Figure 5.23. Research area location (Samani *et al.*, 2018).

the area are covered by lava and young pyroclastic rocks dominated by andesite. The stratigraphic units in the area are as follows
(Figure 5.24):

(1) Cretaceous basic and acidic volcanic rocks, sandstone and spilitic
basalt, calcareous shale (Ksh): these are widespread in the northern and eastern area.
(2) *Miocene pyroclastic conglomerate (MPLvc)*: This unit is more
common in the eastern part of the study area.
(3) Quaternary: the unit includes Quaternary paleoterrace (Qt1),
Quaternary young terrace (Qt2), and Quaternary salt flats
(Qmf), covering a wide area around the Urmia Lake.
(4) River sediments around Ghalechay River.

The Abshir aquifer is an unpressurized aquifer formed by alluvium. The sediments were transported from Sahand pyroclastic rocks
in the northern and northeastern study area through the Ghalechay
River. The alluvium from the middle of the plain to the Urmia Lake
is located on a clay unit. Under the clay, a sand unit forms a confined
brine aquifer. The unit faces toward the lake and contains high saline
water, which is considered to be the residual lake water after the lake
ebbs (Samani and Moghaddam, 2015).

The area inclines from northeast to west and southwest. The highest altitude (about 1320 m) is in the north and east, and the lowest

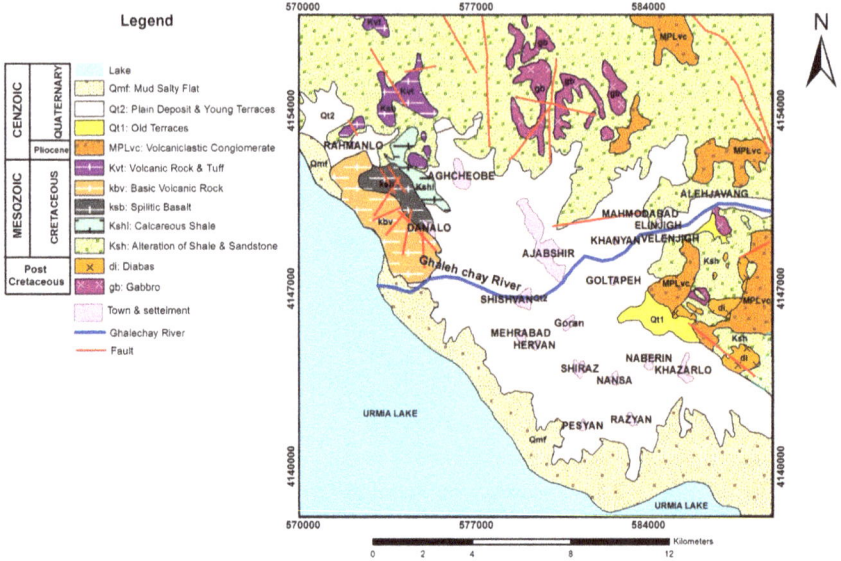

Figure 5.24.   A geological map of the research area (Samani *et al.*, 2018).

altitude (about 1273 m) is in the west. The material and location of bedrock in Ajabshir aquifer were determined by logging and geo- physical data. In the eastern region between the villages of Goltapeh, Goran, and Shiraz to North Nansa, the bedrock consists of igneous rocks. The main components in the north and northeast are the clay, while the main components in the south and central southwest are fine-grained sediments containing lake water. The aquifer bedrock generally follows the surface topography, inclined from northeast to west and southwest. The highest level of bedrock is about 1300 meters in the northeast.

The alluvium in the north and northeast is thinner and gradually thickens in the middle of the aquifer, especially near the middle of the Ghalechay River. The thickest alluvial deposits (about 100 meters) are located in Khanian village and Ajabshir city. The thickness of alluvium gradually decreases away from rivers and toward lakes. Around villages of Shiraz, Nansa, and Nebrin, the alluvial layer is between 60 and 70 meters thick. The average thickness of the allu- vial layer is 50 meters. The largest river in the region is the Ghalechay River, whose source is located in Mount Sahand. The river flows from northeast to southwest and flows into Urmia Lake (Figure 5.24).

The average water level of observation wells in 2013 (Figure 5.25) shows that the general direction of groundwater flow changes with topography, mainly from north and northeast to south and southwest. During this period, the streamline showed the reverse flow from salt flats to plains in some places. It can be seen from the streamline that the Ghalechay River recharges the aquifer in the input and output areas, and parts of it drain the aquifer in the middle section. This flow type is confirmed by data from two hydrological observatories (Yagange upstream and Shishvan downstream). In 2013, the water volume was 59.7 million m^3/year at Yagange Station and 13 million m^3/year at Shishvan Station. During this period, the amount of river water used for irrigation was about 33 million cubic meters. Based on this calculation, the evaporation or infiltration loss in 2013 was 13.7 million m^3/year. Based on the nature of the aquifer and the

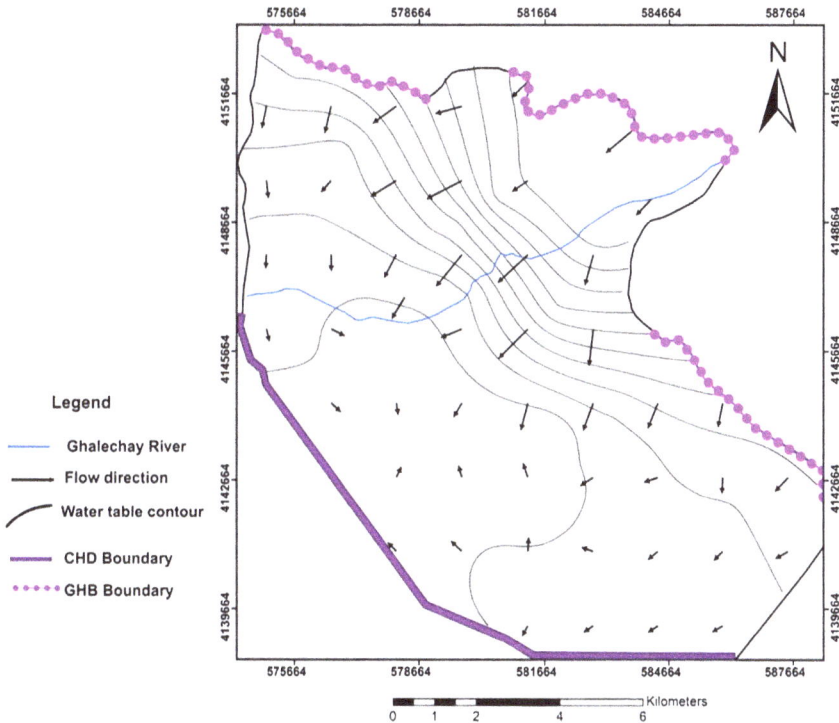

Figure 5.25.   Potential topographic map and flow line in 2013 (Unit: m) (Samani *et al.*, 2018).

area, the surface water infiltration of the Ajabshir aquifer is 8 million m^3/year, assuming 60% of water infiltrating and 40% of water evaporating. Water in aquifers comes from precipitation recharge, river recharge, groundwater flow from mountainous areas around the northern and northeastern plains irrigation return.

The groundwater depth of the aquifer measured in 2013 ranges from 3 m to 35 m. The south and southeast of the aquifer have the lowest and the northwest has the highest water depth. The depth of groundwater is between 4 and 10 m near the salt flats and between 14 and 20 meters in the middle part of the aquifer. There are no wells on the salt flats to measure the groundwater level. However, according to recorded observations, the water levels in these areas are very close to the surface.

The water gradient ranges from 0.001 to 0.02 and gradually decreases from north and northeast to south and southwest. The lowest water gradient is near the salt flats. Based on the pumping test data of 44 pumping wells and 6 exploratory wells, the permeability, horizontal hydraulic conductivity, and water storage coefficient of Ajabshir aquifer are estimated. Due to changes in particle size and aquifer thickness, water permeability varies in different parts of the aquifer, the highest water permeability occurs in the vicinity of Ghalechay River and Ajabshir City, exceeding 1,000 m/day. The water permeability gradually decreases from these areas to the north and the salt flats around Urmia Lake, less than 100 m/day around the salt flats. The water permeability in the middle part of the aquifer and the southern plains around the villages of Nebrin, Nansa, and Shiraz is about 500 m/day. The hydraulic gradient is the largest in the north and northeast, about 20 m/day, and gradually decreases from the north to the southwest. In the salt flats near Urmia Lake, the hydraulic conductivity is the lowest, at about 0.1 m/day. The average hydraulic conductivity of the Ajabshir aquifer is about 12 m/day. The average storage coefficient is estimated at 8%.

Traditionally, model selection criteria such as Akaike Information Criterion (AICC) (Hurvich and Tsai, 1989), Bayesian Information Criterion (BIC) (Rissanen, 1978; Schwarz, 1978), and Kashyap Information Criterion (KIC) (Kashyap, 1982) were used. This is based on the statistical theory. They modify complex models in different ways to achieve a balance between data fitting ability and model complexity. A set of K alternative models (equilibrium and non-equilibrium convection dispersion models) are considered. When $k = 1, \ldots, k$, the model selection criteria are defined as follows:

$$AIC_k = -2\ln[L(\hat{\theta}_k|D] + 2N_k \tag{5.23}$$

$$AIC_{Ck} = -2\ln[L(\hat{\theta}_k|D] + 2N_k + \frac{2N_k(N_k+1)}{N - N_k - 1} \tag{5.24}$$

$$BIC_k = -2\ln[L(\hat{\theta}_k|D] + N_k\ln(N) \tag{5.25}$$

$$KIC = -2\ln[L(\hat{\theta}_k|D] - 2\ln p(\hat{\theta}_k) + N_k \ln\left(\frac{N}{2\pi}\right) + \ln|\overline{F}_k| \tag{5.26}$$

where $\hat{\theta}_k$ is the ML estimate of θk; $-2\ln[L(\hat{\theta}_k|D]$ is the negative log-likelihood (NLL) function; $p(\theta_k)$ is the prior probability of θ_k, $F_k = F_k/N$; N is the number of observations; N_k is the number of parameters. This makes it possible to rewrite KIC_K as:

$$\overline{F}_{k,ij} = \frac{1}{N} F_{k,ij} = -\frac{1}{N} \frac{\delta^2 \ln[L(\hat{\theta}_k|D)]}{\delta\theta_{ki}\delta\theta_{kj}} |\theta_k = \hat{\theta}_k \tag{5.27}$$

$$KIC = -2\ln[L(\hat{\theta}_k|D] - 2\ln p(\hat{\theta}_k) - N_k \ln(2\pi) + \ln|F_k| \tag{5.28}$$

where F_k is the Fisher information matrix; N is the number of observed data; the first term $-2\ln[L(\hat{\theta}_k|D]$, which is common to all criteria, measures goodness of fit between predicted and observed data; the terms containing N_k represent measures of model complexity.

5.2.3.1 *Model construction*

The model Muse is used as the graphical user interface (Harbaugh, 2005), and six 3-D finite-difference numerical models are developed using modular three-dimensional finite-difference ground-water flow model (MODFLOW). The model domain is $134\,\mathrm{km}^2$ and based on the characteristics of the Ajabshir aquifer, six possible conceptual models characterized by alternative geological interpretations, recharge estimates, and boundary condition implementations are compared. To select the transverse boundary of the model based on the natural boundary, the model boundary is extended to the salt flats along the coast of Urmia Lake, and these areas are regarded as a constant water head. According to the available data and hydrogeological conditions, the Ajabshir aquifer is divided into 79 rows, 78 columns, and 1 layer, with a total of 6,162 active units. The cell size of the grid is 200 m in both x and y directions.

Steady-state conditions are used to simulate the water level conditions in 2013. In this year, due to recharge, the groundwater system is close to quasi-steady-state conditions, and the pumping conditions in 2013 are similar to the long-term average conditions. Another reason choosing this year to calibrate is that the available data are reliable and sufficient. Observation wells and pumping wells, boundary conditions, and recharges are all introduced into the model.

5.2.3.2 *Boundary of the numerical model*

The average groundwater level data of 27 observation wells in 2012 was interpolated by fitted surface interpolation method, and used as the initial water head distribution for model. Since the aquifer is unconfined, the surface topography of the model is defined as the highest level. The topographic layer adopts the DEM file of the research area. To build the bedrock layer for the model, geophysical research and exploration logging data are used.

In MODFLOW, well package is used to simulate 574 pumping wells. In this study, in some conceptual models, in addition to the data of pumping wells, well package is also used to input river data (as values of recharge and discharge wells). This is done because sufficient information on river characteristics (riverbed depth, surface water, sediment thickness, etc.) cannot be obtained. Also, this well package is used in some models to replace the input flow boundary (as the value of the recharge well).

The maximum evaporation depth of the Ajabshir aquifer is 4 m. In GIS software, water table depth map is referred to identify areas with a depth of less than 4 m. Only in the southeast of the aquifer near the two observation wells, the water depth is less than 4 m, which is defined as the evaporation area. Another evaporation zone (due to capillary action) is identified near salt flats in the south and southwest of the aquifer. Then, the transpiration rate in 2013 was calculated according to the Thornthwaite method and evaporation percentage.

Based on the groundwater level contours and streamlines in the northwest and southeast of the aquifer in 2013, the flow-free boundary conditions have been determined. Constant head boundary (CHD) is determined in the southern plain and part of the southwest plain bordering Urmia Lake, and it is assumed that the water head of the lake will change with the CHD boundary position. In some areas of the northern and eastern borders, the general head boundary (GHB) is simulated to characterize the inflow and outflow of groundwater at the highland recharge boundary. The general head boundary package can simulate the inflow or outflow of the model according to the difference between the head value in the unit and the specified head value as well as the hydraulic characteristics that determine the ease of flow occurrence. To determine the head level, the northern boundary and the eastern boundary are divided into three segments (Figure 5.25). According to the difference of head and conductance in the north (1300 m and 500 m^2/day) and east parts (1280 m and 300 m^2/day), the real situation of groundwater head simulation is guaranteed. The values of each node are linearly interpolated along with the grid cells. The conductivity in GHB boundary cells is estimated by the hydraulic conductivity value of the boundary region. Head value was adjusted slightly during model calibration and the conductance value of the boundary cells is adjusted by trial and error method. In model number 5, a well package is used instead of GHB while pumping wells are reserved in some cells as input boundary.

The recharge rate of the Ajabshir aquifer was estimated based on precipitation, and the irrigation return water was estimated with recharge package. The amount of recharge varies because of different soil, geology and geological conditions, vegetation characteristics, rainfall intensity, and surface slope. Because little information was available for the amount of recharge, recharge parameters for different areas were calculated during calibration. The Thornthwaite water balance method is used to calculate precipitation infiltration. Based on the existing agricultural wells and irrigation network on the plain, the amount of irrigation return water was estimated by the Blaney–Criddle method. Then recharge was introduced with modeling software as four recharge zones in model number 1. In model number 2, the Ghalechay River was defined with the recharge package, and 3 recharge parameters were defined along the course of the river. In addition, for the area around Ajabshir city, an additional recharge

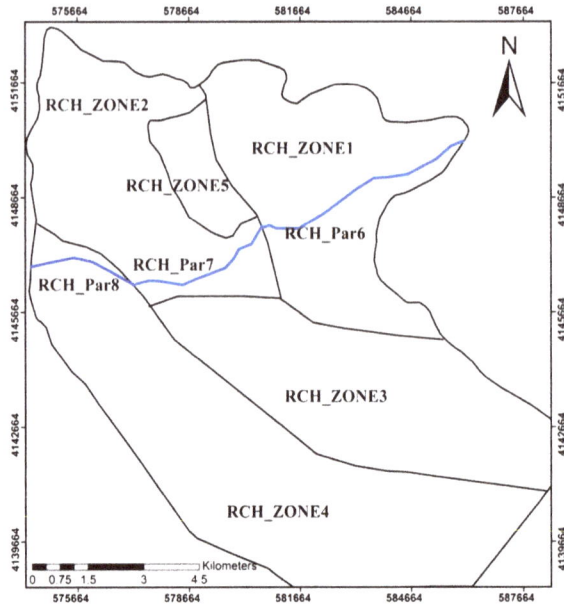

Figure 5.26. The recharge area defined for the modeling software (Samani *et al.*, 2018).

zone was used in this model because of return to the aquifer from municipal sewage and industrial effluents (Figure 5.26).

5.2.3.3 *Hydrogeological zonation and hydraulic parameters*

Hydraulic conductivity data are obtained through pumping tests and combination of aquifer permeability and thickness. For Models 1 and 2, these data are input through interpolation. In Models 3–5, the partition method is used to determine the hydraulic conduction parameters. Data from pumping and exploration logging, hydraulic conductivity estimation standard table, and pumping test are used, so that the research area is divided into four zones. The shape of conductivity zones is calibrated according to geological, lithological, and flow system models and trial and error (Figure 5.27). In Model 6, the hydraulic conductivity parameters are divided into six zones, and further classified based on pumping tests and interpolated hydraulic conductivity maps (Figure 5.27).

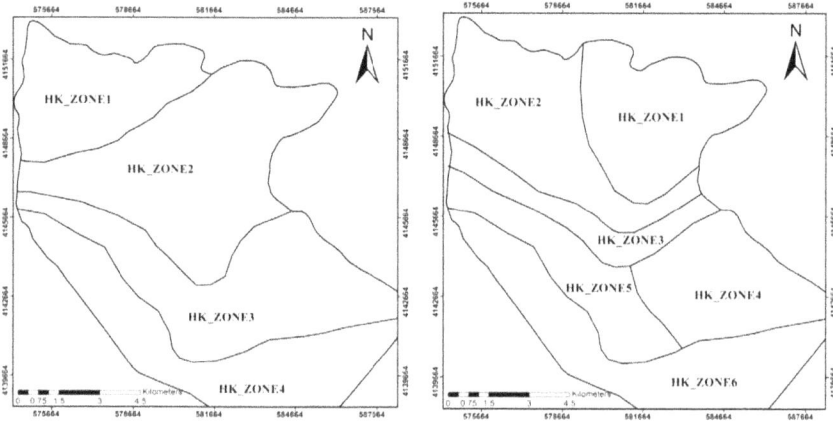

Figure 5.27. The hydraulic conductivity zonation defined by modeling software: Model 3, Model 4, and Model 5 are four regions and Model 6 is six regions (Samani *et al.*, 2018).

Table 5.12. Statistics for calibration and validation of six alternative conceptual models (Samani *et al.*, 2018).

Model	Model 1	Model 2	Model 3	Model 4	Model 5	Model 6
SSR (m^2)	15.28	13.052	13.752	9.648	11.893	8.021
RMSE calibration	0.751	0.695	0.714	0.598	0.664	0.545
Residual mean (m)	−0.025	0.030	0.067	0.026	0.002	−0.022
Abs.Res Mean (m)	0.625	0.564	0.628	0.513	0.568	0.387
Res.Std.Dev (m)	0.765	0.708	0.724	0.609	0676	0.471
Abs.Res mean divided by range	0.012	0.011	0.013	0.010	0.011	0.008
Std.Dev divided by range	0.015	0.014	0.014	0.012	0.014	0.009
Model ranking according RMSE calibration	6	4	5	2	3	1
RMSE validation	0.774	0.832	0.734	0.819	0.681	0.609
Model ranking according RMSE calibration	4	6	3	5	2	1

The performance of each model was evaluated by the goodness of fit between observed and simulated hydraulic heads proposed by ESI (2007): mean absolute residual (MAR) and residual standard

deviation divided by range of heads less than 10%. The difference between the maximum and the minimum water level (head
range) measured in the Azabhir aquifer is 50 m. Table 5.12 summarizes the sum of squared residuals (SSR), root mean square error
(RMSE), mean residual error, MAR, and residual standard deviation of the six conceptual models. The MAR and residual standard
deviation divided by head range of all models are between 1 and 2%
(Table 5.12), so the statistical results for quality of data fit are below
the calibration target.

5.2.3.4 *Influence of complexity on uncertainty of groundwater modeling*

To determine the uncertainty caused by the complexity of the model,
alternative conceptual models with different degrees of complexity
are created. The first (easiest) model (number 1) is designed with
five parameters. The complexity is increasing by adding parameters to subsequent models: Models 2 and 3 have 10 parameters,
Models 4 and 5 have 13 parameters, and Model 6 has 15 parameters. Through the following analysis, it is determined whether the
most sophisticated 15-parameter model performs best or the simplest
6-parameter model performs best enough to describe the groundwater flow in Ajabshir aquifer.

In order to examine the complexity according to model probability, method proposed by Nettasana (2012) is used. Table 5.13 shows
the results for prior model probability calculated. The highest prior
model probability was found for model 6, and this probability is used
to calculate posterior model probability with different model selection criteria (AIC, AICc, BIC, KIC).

In the AIC method, the simplest model (1) obtains the highest
model probability, and the most complex model (6) obtains the second highest model probability. It is worth noting that this method
could help to observe the probability distribution between models.
Therefore, using the AIC and model averaging method may obtain
a certain prediction result effectively.

AICC and BIC methods yield no the probability distribution in
each alternative model, and both methods identify that the probability of model 1 is the largest. These two methods show that the
complex model with many parameters has higher uncertainty, while

Table 5.13. Parameter uncertainty and parameter estimation of five parameters common to alternative conceptual models (Samani *et al.*, 2018).

Model parameter	Model 1	Model 2	Model 3 RCH-Par1	Model 4	Model 5	Model 6
Parameter uncertainty	8.26E-05	2.43E-04	2.25E-04	3.32E-04	1.73E-04	3.90E-04
Parameter estimation	2.64E-03	1.95E-03	4.24E-03	2.97E-03	1.61E-03	2.27E-03
Parameter			RCH-Par2			
Parameter uncertainty	3.85E-05	5.88E-05	1.00E-04	1.37E-04	4.95E-05	9.35E-05
Parameter estimation	1.01E-03	2.91E-04	4.10E-04	2.23E-04	7.58E-04	1.00E-06
Parameter			RCH-Par3			
Parameter uncertainty	3.69E-05	7.41E-05	1.10E-04	9.50E-05	7.21E-05	9.45E-05
Parameter estimation	2.81E-04	4.42E-04	1.03E-04	6.44E-05	5.51E-04	2.35E-04
Parameter			RCH-Par4			
Parameter uncertainty	3.14E-05	1.20E-04	8.34E-04	5.87E-04	1.80E-04	5.29E-04
Parameter estimation	5.91E-04	1.20E-03	3.01E-03	2.70E-03	1.00E-06	1.11E-03
Parameter			EVT-Par1			
Parameter uncertainty	5.81E-05	1.17E-04	5.21E-04	3.00E-04	1.19E-04	3.94E-04
Parameter estimation	3.90E-04	9.7E-04	1.84E-03	1.34E-03	1.60E-04	5.19E-04

the simple model with few parameters has lower uncertainty. So that, a simple model with few parameters is sufficient to describe Ajabshir aquifer, and over-parameterization will lead to uncertainty. When AICc and BIC methods are used for prediction, the model averaging method is not needed.The KIC method is the most complete method to evaluate model complexity, and is the only one able to discriminate between models based not only on their goodness of fit to observed data and the number of parameters, but also on the quality of the available data and of the parameter estimates. When KIC method is used, both simple and sophisticated models show low probability and high uncertainty. This method uses the sensitivity of model parameters (Fisher information matrix) to select the model with average complexity. This method shows that neither simple nor complex models can well describe the characteristics of aquifers and will cause uncertainties. Therefore, choosing the model with the optimal parameters can avoid the uncertainty caused by the complexity of the model. When predicting using KIC method, the model average method is not need any more.

It should be noted that the probability of Models 2, 4, 5, and 6 is zero when AICc and KIC methods are used. This result reflects that the concept definitions used in these models are inappropriate. Before calculating the probability of the model, six conceptual models are defined without making a prior assumption on the accuracy of the basic definitions. However, the results of model probability comparison reject the trueness of the conceptualization of Models 2, 4, 5, and 6. Therefore, defining recharge package instead of the river in Models 2 and 4, defining river instead of the general head boundary in model 5, and changing the zoning of conductivity in Model 6 will lead to the wrong conceptual model, resulting in zero model probability.

Model 3 has the same degree of complexity as Model 2 and has a smaller Fisher term, so the KIC value is also smaller. Therefore, it is speculated that the quality of data and parameter estimation play a greater role in model determination rather than the complexity of the model. This can explain why the KIC method is more inclined to use moderately complex model 3 in parameter estimation.

According to the maximum likelihood theory, the mapping of each parameter shows that the parameters of each calibration model follow a normal distribution, the mean value is the estimated value of the parameters, and the standard deviation is the variance of each parameter. This type of graph is also called a posterior parameter distribution (Ye *et al.*, 2010). From the figure, we can identify the model with higher certainty parameters. The conceptual models of the Ajabshir aquifer compared has five common parameters: RCH_Par1, RCH_Par2, RCH_Par3, RCH_Par4, and EVT_Par1. The values used to draw the posterior parameter distribution are shown in Table 5.13 and Figure 5.28. The lowest posterior parameter probability (i.e., the greatest uncertainty) in parameter estimates were seen in Models 3, 4 and 6, and the highest posterior parameter probability was seen in Model 1. The posterior parameter probability distribution of Model 1 is significantly different from other models,

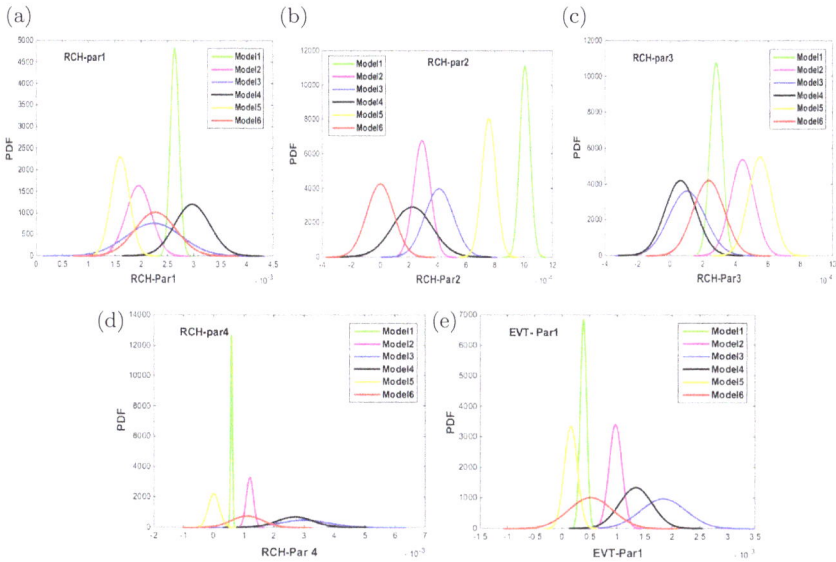

Figure 5.28.   Posterior distribution of recharge and evaporation parameters in models (Samani *et al.*, 2018).

which can be explained as the significant difference in the number of parameters (6 parameters versus 10, 13, or 15 parameters).

It is noted that parameter sensitivity is very important in the KIC method, so this method sometimes prefers models with high parameter uncertainty (Ye *et al.*, 2008, 2010). This is evident in Figure 5.28, where the KIC method selects Model 3 as the best model, although the uncertainty of parameters in this model is greater than that of other models.

5.2.3.5   *Conclusion*

The six groundwater flow models of the Ajabshir aquifer were compared using the data of 2013, and the performance of the model was validated by data of 2014. To determine the uncertainty caused by model complexity, six models are divided into four complexity levels, with, 6, 10, 13, and 15 parameters. The complexity of the two pairs of models is the same (Models 2 and 3, Model 4, and Model 5). The models are scored according to RMSE and SSR. The model with 15 parameters is ranked best, and the model with 6 parameters is ranked worst.

According to the model selection criteria, the prior model probability calculated by the model calibration results is used to calculate the posterior model probability. Model 1 (the simplest model) has the highest model probability in AIC, AICc, and BIC. The comparison with these methods shows that a simple model with fewer parameters is suitable for describing the groundwater dynamics of Ajabshir aquifer while over parameterization will lead to greater uncertainty.

KIC method can distinguish models not only based on the goodness of fit to observational data and the number of parameters, but also based on the quality of data and parameter estimation by Fisher information matrix. This method chooses Model 3 as the best model with the greatest probability.

5.2.4 Variable model simulation of groundwater flow in the glacial aquifer system in northeastern Wisconsin

This study was carried out by Juckem *et al.* (2017). The National Water-Quality Assessment (NAWQA) is responsible for mapping the intrinsic sensitivity of groundwater in glacier aquifers connected to the United States. Eberts *et al.* (2013) define intrinsic sensitivity as "a measure of the ease with which pollutants in water enter and pass through aquifers". It is the characteristic of aquifers and overlying substances and has nothing to do with the characteristics of sources of pollutants. "Inherent vulnerability to contamination is an important indicator of groundwater age, groundwater with younger ages indicating recent recharge or rapid movement and therefore potentially being susceptible to surface activity; older water is generally more susceptible to natural pollutant due to geochemical processes. Therefore, understanding the rate and spatial pattern of groundwater recharge, combined with the velocity of groundwater flow in the glacier aquifer system, is important to estimate the inherent sensitivity of the glacier aquifer.

Due to the complexity and scale of glacial aquifers, which extend from the east coast to the west coast and cover parts of many northern states of the United States (Figure 5.29), meta-modeling method is envisaged to map the inherent sensitivity of aquifers. The metamodel is a statistical model trained according to the output of a process-based computer model. Therefore, it is freed from many constraints of a process-based complex groundwater flow model, such as long operation time and spatial range (Fienen *et al.*, 2015). Inputs to the meta-model may include important hydrogeological properties, such as aquifer thickness, groundwater recharge, and glacier lithology. The training or calibration of the meta-model is expected to include comparing the meta-model output with the simulated age distribution of process-based groundwater flow and flow transport models (such as MODFLOW and MODPATH) (a particle-tracking model for MODFLOW). However, more work needs to be done to understand different interpretation methods and how methods of integrating surface data sets into the three-dimensional groundwater modeling framework will affect the simulated age distribution that will be used to train meta-models.

This case describes the construction of three groundwater flow models and calibration at different levels of complexity to simulate groundwater flow in northeastern and surrounding of the Fox-Wolf-Peshtigo (FWP), Wisconsin (Figure 5.29). More attention was paid to the complexity difference of the three-level models of the glacier aquifer water level simulation system. These models also form a framework for future evaluation, using model complexity to simulate groundwater age distribution.

5.2.4.1 *Description of research area*

The Fox River and its largest tributary, the Wolf River, are part of the Lake Michigan watershed. The Fox and Wolf rivers flow from the source area along the Great Lakes–Mississippi subcontinent watershed, pass through Lake Winnebago, and finally discharge to Green Bay. The Oconto and Peshtigo rivers, along with the Duck Creek and Pensaukee rivers, discharge water from the northern part of Winnibago and also discharge to Green Bay. From the mouth of Green Bay to the source tributary, the Fox River basin covers an area of 64,000 square miles, while the rest of the watershed in the study area covers an area of 27,000 km^2. In this work, the scope of the model also includes tributaries of the Wisconsin River, including Plover River, Eau Claire River, and Pine River, as well as several streams in western Waupaca. Lakes and wetlands are common throughout the study area, especially in the northern forested part of the study area.

Figure 5.29.   Location of the Fox-Wolf-Peshtigo and the connected glacier aquifer system, Wisconsin (Juckem *et al.*, 2017).

The landscape is closely tied to glacial landforms, with glacial moraines forming many watershed divides. Its highest point is Sugar Shrub Hill at the source of the Wolf River, about 1,910 feet above the 1988 North American Vertical Datum (NAVD 88), and its lowest elevation is Lake Michigan, about 580 feet above NAVD 88.

5.2.4.2 *Conceptual model of the groundwater system*

The development of groundwater flow models is based on the conceptualization of groundwater hydrogeological systems and their interactions with surface water and groundwater wells. Groundwater–surface water exchange is expected to be concentrated in shallow glacier aquifer systems (Feinstein *et al.*, 2016). Groundwater is generally discharged to surface water bodies, and only partial surface water is recycled back to aquifers. Groundwater is extracted from coarse glacial sediments (mainly in the west) and bedrock aquifers with increased thickness in the east.

In the FWP, groundwater flows in two main aquifers, the surface glacier aquifer system and the lower bedrock aquifer system. In the north and west of the study area, glacier aquifer directly overlies impermeable crystalline bedrock, is the main aquifer. In the south and east of the study area, permeable sandstone and dolomite aquifers provide substantial storage and transmissivity, whereas the glacial deposits usually contain more fine-grained materials. The bedrock surface separating the glacier aquifer from the bedrock aquifer is usually irregular; therefore, the covered glacial sediments show high variability in thickness.

The glacial materials in the FWP study area show different lithologies due to sedimentary environments (Lineback *et al.*, 1983; Farrand *et al.*, 1984). Although the written description of the surface deposits on the map is generally limited to the upper few meters (Lineback *et al.*, 1983; Farrand *et al.*, 1984), the lithology analysis of the well construction report shows little variation with depth compared to horizontal transitions in lithology. Therefore, it is expected that the map of surface glacier units in the FWP study area will still be useful for understanding the geological characteristics of glacier aquifer systems at all depths. This assumption may not apply to other parts of the glacier aquifer system in the United

States, where deep glacier deposits are associated with many complex glacier advances and retreats.

Sandstone and dolomite overlay the crystal bedrock, forming a ridge along with the Wisconsin Arch, a southern extension of the Precambrian Canadian Shield, that forms a peninsula shape between the Michigan Basin to the east, the Illinois Basin to the south, and Hollendale Embayment (Mossler, 1992) to the west. In the FWP study area, these sedimentary bedrock units dip eastward and increase the thickness toward the Michigan Basin. Considering that the project focuses on glacier aquifer, Feinstein *et al.* (2016) used a simplified double bedrock aquifer representation. That is, the bedrock strata are simplified to the upper unconfined bedrock aquifer and the lower confined bedrock aquifer. Under this concept, there are no single regional continuous strata confining unit to separate the two bedrock systems. Instead, the uppermost important confining unit is used to separate the upper system from the lower system, and the confining unit is considered to be part of the lower confined system. Although an important source of water supply in bedrock aquifers is the main target that has been investigated in advance, bedrock aquifers are included in this study mainly considering their potential impact on aquifer permeability and thickness and therefore their potential impact on groundwater age in glacial aquifer. In addition, the large cone of depression that was caused by historical and contemporary well withdrawal in the bedrock aquifers along the Lower Fox River valley was depicted by Feinstein *et al.* (2010) to divert some groundwater from the overlying glacial aquifer into the bedrock system, which is expected to locally affect simulated groundwater ages.

5.2.4.3 *Establishment of the groundwater system model*

Model construction involves generating input arrays (gridded data) describing the spatial distribution of aquifer hydraulic conductivity, recharge, top and bottom elevations for each model layer, and mapping curvilinear surface water bodies into square cells, and applying well production rates to cells and layers based on well construction information. The groundwater flow system adopts the groundwater flow model MODFLOW-NWT, a Newton-Raphson formulation for MODFLOW-2005 (Niswonger *et al.*, 2011) to solve the groundwater flow equation by an upstream-weighted block-centered finite-difference method (Harbaugh, 2005; Niswonger *et al.*, 2011). The MODFLOW-NWT model is designed in part to address the challenge of simulating thin unconfined aquifers (common in glacier regions), which are vulnerable to oscillating dry cell problems during iterative solutions. When the water level falls below the bottom layer, dry cells will occur. MODFLOW-NWT minimizes this problem by assigning a very small minimum thickness to each cell. For the final solution, MODFLOW-NWT assigns a user-specified value (-999 for the FWP model) to the simulated head in the cell, with a thickness less than or equal to the specified minimum thickness, which is set to the default value of 0.00001 feet.

Three MODFLOW models are developed to simulate the FWP research area under different model complexity and vertical discretization. However, the input to the model is shared by three models, including the following: (1) the initial recharge estimated by a Soil Water Balance project (Westenbroek *et al.*, 2010); (2) all boundary packages simulating surface water characteristics; (3) the adjusted bedrock surface elevation representing the base of the glacial aquifer system; (4) simplified glacial categories based on the Quaternary Atlas (Lineback *et al.*, 1983; Flanders *et al.*, 1984). The three models differ by the following: (1) the number of layers used to represent the glacial aquifer system; (2) whether the bedrock aquifer is simulated or considered to represent a lower no-flow boundary; (3) the level of heterogeneity in the hydraulic conductivity fields used to represent glacial aquifer deposits. Of the three models, two simulate the bedrock aquifer and include additional withdrawal wells and target water levels for wells penetrating the bedrock aquifer. These additional production wells and water level targets are not included in the one-layer model that ignores the simulation of bedrock aquifers.

Model construction includes these key points, which are briefly stated as follows.

The MODFLOW model is designed with cells spanning 1,000 feet per side, including 930 rows of 650 columns of cells per layer (604,500 cells per layer), covering 21,680 square miles. In the three versions of the FWP model, the vertical discretization of the three hydrological stratigraphic elements (glacial aquifer, upper unconfined bedrock aquifer, and lower confined bedrock aquifer) identified in the conceptual model is different.

Precambrian crystalline rocks, only supply water when there are no other available resources, are no-flow boundaries at the bottom of glacier and bedrock aquifer systems. In addition, the Wisconsin River and Lake Michigan are effective hydrological boundaries along the western and eastern edges of the model, as their water levels are the lowest in their respective regions, so little or no water is expected to flow from below them through the glacial aquifer system.

The Lakes of Michigan and the Rivers of Wisconsin are simulated using general head boundary (GHB) package (Harbaugh *et al.*, 2000), in which the water levels of these large features are specified. Rivers within the model domain are simulated by streamflow-routing (SFR2) package (Niswonger and Prudic, 2005), which routed water from the upper reaches to the lower reaches to accumulate flow and solved water levels in the river. The Unsaturated-Zone Flow (UZF1) package is used as a flexible tool to simulate aquifer recharge or groundwater discharge into lakes and wetlands. Using the Soil-Water-Balance (SWB) program described by Westenbroek *et al.* (2010) to compute the average rate of deep infiltration or potential groundwater recharge for a long time (approximately 1980–2012).

Three MODFLOW models are built to simulate different degrees of aquifer complexity. The first model is a 1-layer model of glacial aquifer sediments that uses a refined bedrock surface to represent the base of the groundwater flow system (bedrock is not simulated). The second and third models use three layers to represent glacial aquifer sediments and two underlying layers to represent unconfined and confined bedrock aquifers.

In addition to adjusting the number of layers used to represent aquifers, the levels of complexity used to represent glacial sediment permeability vary in order to evaluate the heterogeneity of the simulated age distribution in future studies. Both the 1-layer model and one of the two 5-layer models used this simplified glacier classification map to simulate internally homogenous hydraulic conduction zones. In the second step, cell–cell heterogeneity is generated in each glacier category by interpolating coarse fractions estimated from lithology descriptions in well logs. The resulting heterogeneous hydraulic conductivity field is applied to the second 5-layer model, which is subsequently called the 5-layer heterogeneous model.

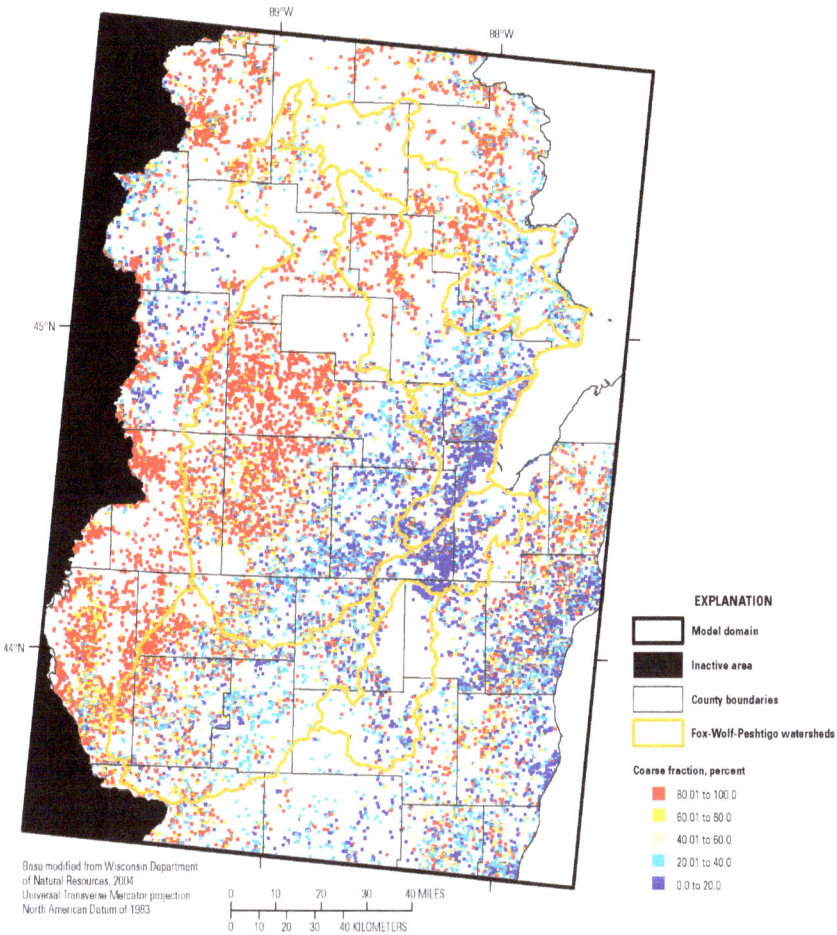

Figure 5.30.   Composite coarse component for the model unit containing lithological logging data from the standardized lithology database by Bayless and
others (Juckem *et al.*, 2017).

All three versions of the FWP model use the same glacier category
map. A single hydraulic conductivity value is assigned to each glacier
category in the zoned model, whereas the coarse fractions distribution (Figure 5.30) are used to calculate the heterogeneous hydraulic
conductivity field for the 5-layer heterogeneous model. The coarse

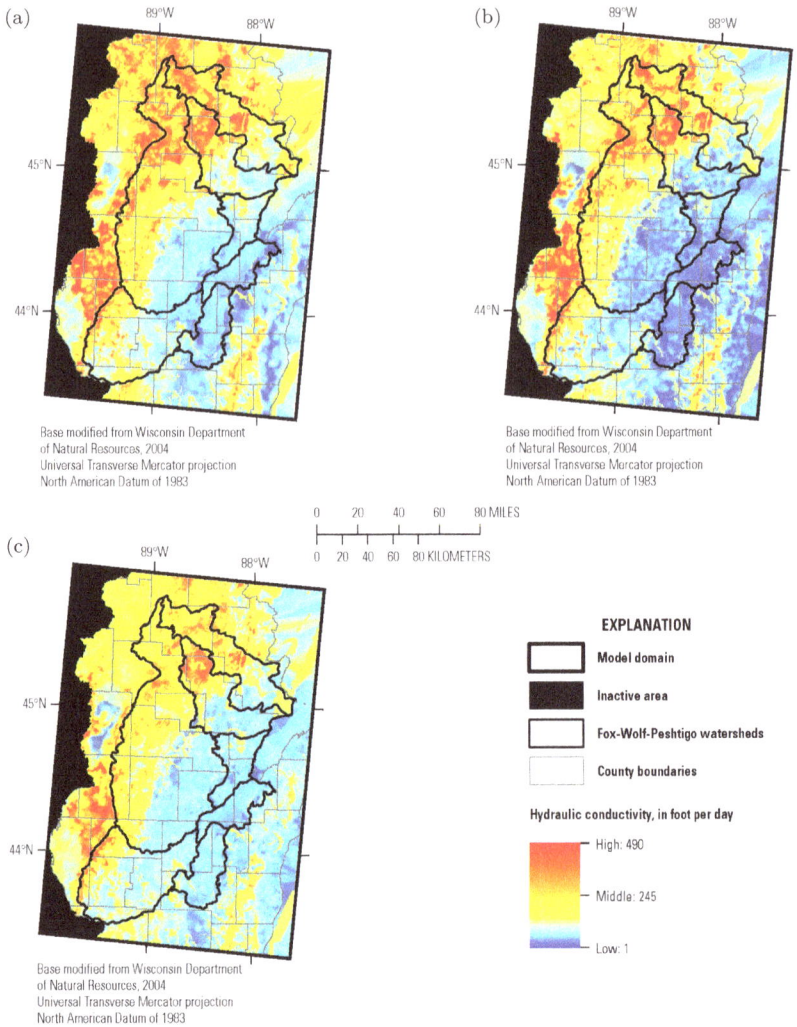

Figure 5.31. (a) Layer 1 to Layer 5 heterogeneous model glacier sediments coarse fractions layer by layer; (b) Floor 2; (c) Floor 3 (Juckem *et al.*, 2017).

fractional values of each cell in the model layer are calculated by spatially interpolating by kriging values of the cells (Figure 5.31). Similar to the bottom elevation of the bedrock layer, the horizontal and vertical hydraulic conductivity used in the 5-layer model to represent the bedrock aquifer is derived from the corresponding bedrock layer in the "intermediate" Lake Michigan Basin (LMB) model described by Feinstein *et al.* (2016).

5.2.4.4 *Model calibration*

Model calibration is the process of adjusting the initial input values (e.g., hydraulic conductivity) to improve the matching of the analog outputs (water level and base flow) with the corresponding field measurements or targets. From the simplest to the most complex, each FWP model is independently calibrated using Parameter Estimation Simulation Technique (PEST) (Doherty, 2010), which seeks to minimize the difference between the simulated and measured target values by adjusting the parameter values within a user-specified limits. More specifically, during calibration, PEST minimizes the sum of the squared-weighted between the simulated and the measured target value (PHI).

The FWP model is calibrated to 200 long-term (1970–99) base flow targets and thousands of water level targets. All three models are calibrated to the same target set, except for the 5-layer models which are also calibrated to the water level of the wells open from the bedrock aquifer. The 1-layer model is limited to the water level targets open only to the glacial aquifer that are not affected by the vertical gradient. All three FWP models are calibrated in a similar way, that the same set of parameters are estimated in the calibration step, using a ratio or fixing at initial value. However, the 5-layer models have additional parameters, representing the ratio of horizontal to vertical water hydraulic conductivity, and parameters representing the bedrock aquifer, which are not required in the 1-layer model.

Direct comparisons of model calibrations are best assessed using only glacial aquifer system wells and base flow targets. Table 5.14 presents summary statistics for glacial aquifer water levels, calibration statistics for base flow targets are presented in Table 5.15.

Table 5.14 shows that as the complexity of each model increases, the simulated water level of the glacial aquifer system generally improves. That is, the mean absolute error, root mean square error (RMSE), and PHI decrease when the 1-layer model is compared with the 5-layer zoned model, and when the 5-layer zoned model is compared with the 5-layer heterogeneous model. Increasing the heterogeneity of the glacial aquifer system appears to improve water

Table 5.14. Summary statistics for each model using water-level targets in the glacier aquifer system (Juckem *et al.*, 2017).

Model	Number of water-level targets in the glacial aquifer	Mean error for water levels (ft)	Mean absolute error for water levels (ft)	RMSE for water levels (ft)	Range in measured water levels (ft)	RMSE/range	Sum of squared weighted residuals (PHI in PEST) for glacial well targets	Total PHI (percent)
1-layer model	4,400	0.18	11.8	17.6	1,158	0.015	148,992	54
5-layer zoned model	4,400	1.70	11.5	17.3	1,158	0.015	134,057	32
5-layer heterogeneous model	4,400	0.78	9.92	15.1	1,158	0.013	97,346	25

Table 5.15. Summary statistics for each model using all base-flow targets (Juckem *et al.*, 2017).

Model	Number of base-flows targets	Mean error for base flows (ft3/s)	Mean absolute error for base flows (ft3/s)	RMSE for base flows (ft3/s)	Range in measured base flows (ft3/s)	RMSE/ range	Sum of squared weighted residuals (PHI in PEST) for base-flow targets	Total PHI (percent)
1-layer model	200	2.8	15.8	38.6	1,460	0.026	129,269	46
5-layer zoned model	200	3.5	15.6	38.7	1,460	0.026	127,794	30
5-layer heterogeneous model	200	2.8	15.3	36.7	1,460	0.025	121,401	32

level residuals more than increasing the number of layers. However, it is challenging to directly compare the 5-layer zoned model with the 1-layer zoned model because of the variation of multiple factors between these models. Some factors include the following: (1) dividing glacial aquifer into 3 layers, (2) adding bedrock layers, (3) including 119 glacial targets that appear to be affected by vertical gradients, and (4) including water level targets in bedrock aquifers. Item 1 and 2 will provide more detail and realism to improve water level simulations in glacial aquifer system, however, the addition of new targets will create logical competition with the original targets used to calibrate the 1-layer model and may reduce the match with measured water levels for 4400 wells summarized in Table 5.14.

Therefore, if each model is calibrated using only the 4400 targets, it is difficult to determine whether the improvement in summary statistics observed between the 1-layer and 5-layer zoned models might have been greater. Regardless, the two 5-layer models perform better on the 4400 glacier-level targets than the 1-layer model, despite being calibrated to other targets. Summary statistics for base flow targets (Table 5.15) show that modest reductions in both the mean absolute error and PHI values when comparing the 1-layer model with the 5-layer zoned model, and also when comparing the 5-layer zoned model with the 5-layer heterogeneous model. However, the RMSE increases slightly from 1-layer to 5-layer zoned models. Thus, the results show that increasing complexity generally has less impact on the overall fit of the base flow target than the water level target.

Further insight into the impact of increasing complexity on calibration metrics can be obtained by directly comparing targets among models. The changes in absolute error for each glacial well and base flow target were compared by subtracting the results of the less complex model from the more complex model and are summarized in Tables 5.16 and 5.17. The results presented in Table 5.16 show that a larger proportion of water level targets (59%) showed improvement (lower absolute errors) when adding heterogeneous hydraulic conductivity to the model compared to adding layers (51%). Furthermore, the magnitude of improvement (4.6 ft) and degradation (2.9 ft) are greater when adding heterogeneity than when adding layers to the model (1.9 and 1.3 ft). The results show that adding heterogeneity has more effect (larger magnitude) on simulated water level targets,

Table 5.16. Change in absolute error for water-level targets in the glacial aquifer system (Juckem *et al.*, 2017).

Models compared	Number of water-level targets in the glacial aquifer	Number of improved water-level targets in the glacial aquifer	Improved water-level targets in the glacial aquifer (percent)	Average reduction in absolute error for improved targets(ft)	Average increase in absolute error for degraded targets(ft)
5-layer zoned model versus 1-layer model	4,400	2,272	51	1.9	1.3
5-layer heterogeneous model versus 5-layer zoned model	4,400	2,617	59	4.6	2.9

Table 5.17. Change in absolute error for base-flow targets (Juckem *et al.*, 2017).

Model compared	Number of base-flows targets	Number of improved water-level targets in the glacial aquifer	Improved water-level targets in the glacial aquifer (percent)	Average reduction in absolute error for improved targets(ft3/s)	Average increase in absolute error for degraded targets(ft3/s)
5-layer zoned model versus 1-layer model	200	114	57	1.0	0.95
5-layer heterogeneous model versus 5-layer zoned model	200	103	51	3.8	3.6

both in terms of improvement and degradation than adding layers. This result is not surprising considering that 119 targets suspected of being affected by vertical gradients are removed from the 1-layer model and, therefore, removed from this analysis.

Since the additional estimated parameters of the 5-layer zoned model are related to the vertical anisotropy and permeability of the bedrock compared to the 1-layer model, parameters not expected to have a strong effect on water levels in a largely Dupuit-Forchheimer dominated flow system (little resistance to vertical flow along predominantly horizontal flow paths). Instead, for the heterogeneous model, additional 6 horizontal hydraulic conductivity parameter values are calibrated. Although any increase in parameter flexibility logically provides greater flexibility to match the calibration target, it turns out that the system is largely dominated by horizontal flow; thus, the water level target responded more to complexity associated with horizontal hydraulic conductivity than vertical hydraulic conductivity.

Comparison of base flow targets between models of increasing complexity (Table 5.17) is less clear, with a larger percent of baseflow targets indicating improvement due to the adding layers (57%) than increasing heterogeneity (51%). However, the magnitude of the change in base flow is greater with the addition of heterogeneity than with the addition of layers. By spatially plotting the variation of residuals for the 4,400 targets summarized (Figure 5.32), it is convenient to determine where and how the additional complexity can improve the simulated water levels. Changes in residuals are computed using the absolute error for each target, moving from a simpler model to a more complex model (1 to 5 layers; zoned to heterogeneous). The results show that using a 5-layer zoning model with three layers instead of one (Figure 5.32a) to simulate a glacial aquifer system produced small changes (less than 5 feet difference) for most targets, which is consistent with the results in Table 5.16.

Nonetheless, the results show that the greatest improvements are within the loamy till and fine-stratified glacial categories, especially in the source areas where the vertical gradient is expected to be greatest. In fact, many of the 119 targets that are removed from the 1-layer model calibration because of perceived vertical gradients are within the fine-stratified glacial category and likely would have exhibited a similar improvement with the addition of model layers. Most targets in the coarse grained glacier category do not be affected by the addition of model layers.

Adding heterogeneity in the 5-layer model appeared to improve the simulated water level (by at least 5 feet) for a greater number of glacial targets (Figure 5.32b) than simply adding layers (Figure 5.32a). The results show that the greatest improvements were within the coarse-stratified and sandy till glacial categories, as well as for some wells in the fine-stratified category. This result is consistent

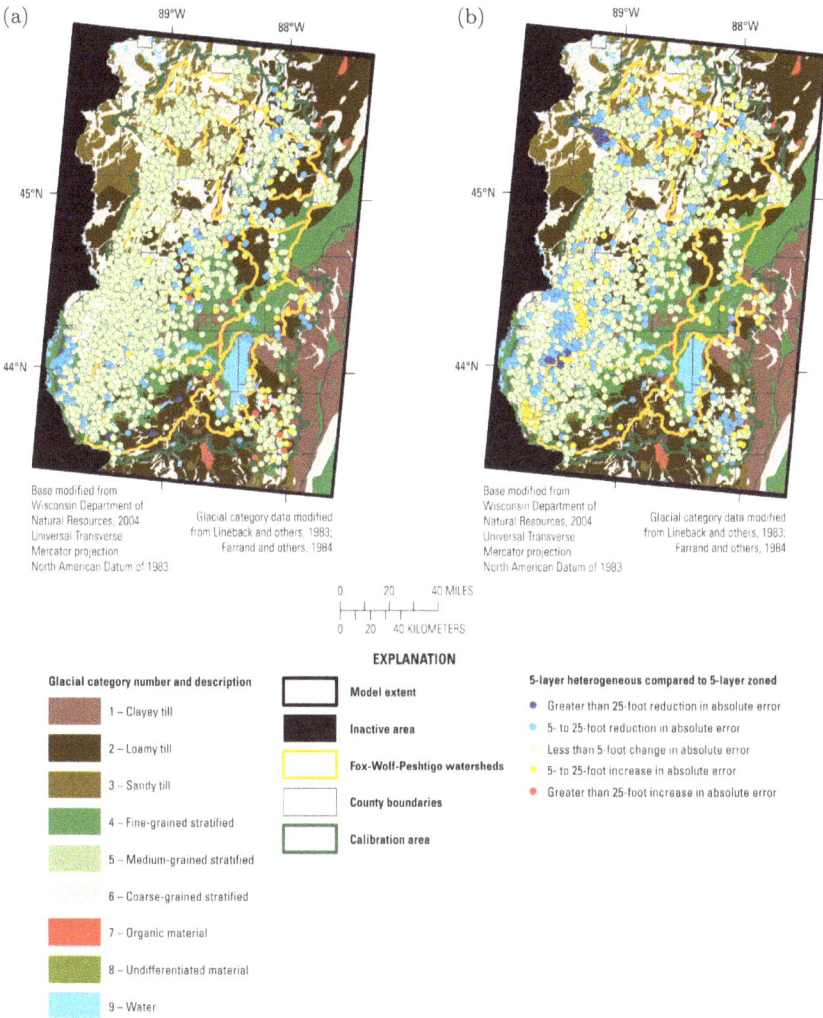

Figure 5.32. Change in water-level target residuals between (a) the 5-layer zoned model and the 1-layer model and (b) the 5-layer heterogeneous model and the 5-layer zoned model (Juckem *et al.*, 2017).

with the sensitivity of the target to these glacier categories. These three glacier types cover most of the model domain, suggesting that heterogeneity may be an important factor controlling water levels in much of the FWP study area. Improvements in the base flow target were roughly similar to the water level targets, with a larger change observed when increasing the heterogeneity compared to simply increasing the number of layers (Table 5.17). However, the improvement is small compared to the water level target, and spatial inferences of the causes are complicated by the large accumulating areas that affect base-flow targets; therefore, the improved base flow residual plot is not shown for brevity.

5.2.4.5 *Limitations of groundwater flow models*

Like all models, the groundwater flow model described in this report is a simplification of the physical system and has corresponding restrictions on the accuracy of the model and how to use the model. For example, the model calibration process focuses on steady-state water level and base-flow targets to estimate hydrogeological properties. Therefore, the model cannot reliably simulate seasonal water level fluctuations. Also, due to the discretization of model cells (1000-foot wide cells), the conditions within each cell (e.g., groundwater level) are reduced to an average value of the cells. Although the size of this cell is expected to be sufficient to meet the simulation objectives described in this report, a new analysis requiring finer spatial details will benefit from model improvement. Also, the local geological complexity not included in the model, the performance of the model may vary at different location; in other words, the calibrated hydraulic conductivity fields of the 5-layer heterogeneous model were designed to incorporate heterogeneity as a function of coarse fraction described in lithologic logs. When the local lithologic logging lacks sufficient accuracy or density, there may be additional heterogeneity on the local scale. Moreover, the method divides distant areas into glacial categories according to extensive lithological descriptions. For areas where this correlation does not match, the calibrated parameters may not well reflect the local geological conditions. Similarly, the regional design and calibration of the model further limit the local scale accuracy of model parameters and model results. That is, reference to the parameter values used in these models, such as hydraulic conductivity and recharge capability should focus on the average level of the region or the region rather than the local extreme situation.

The purpose of this set of models is to provide a basis for future analysis of groundwater age distribution in glacier aquifer systems and evaluation of some impacts of complexity on simulated water levels. Applying these models for other purposes, such as assessing the impact of groundwater extraction on surface water characteristics, may be of limited value and will benefit from further assessment of calibration parameters, grid resolution, and surface water conductance parameters that affect groundwater/surface water interaction. Similarly, problems focused on the assessment of the impact of groundwater extraction on surface water characteristics, may benefit from simulating specific water bodies of interest with more advanced packages, such as the Lake Package (Merritt and Konikow, 2000). This work also focuses on glacial aquifer systems and greatly simplifies bedrock aquifers; interested readers can refer to other resources, such as Feinstein *et al.* (2010) as a starting point for assessing groundwater in bedrock aquifers. Also, the potential application of models that are important for groundwater flow rates (e.g., mapping time-referenced contribution areas to well, such as "10-year area contributing recharge to a well") may benefit from the evaluation of porosity and groundwater age tracer concentrations. Local grid refinement or use of unstructured grid (Panday *et al.*, 2013) may be guaranteed to calculate more accurate groundwater flow particle curve paths or simulation of related particle travel times.

Finally, this case does not consider the influence of parameter uncertainty when comparing different degrees of model complexity. In other words, the model calibration process often produces non-unique parameters, so multiple groups of parameters can produce similar calibration metrics. Parameter estimation simulation technique (PEST) (Doherty, 2010) provides linear and nonlinear tools to evaluate the impact of this uncertainty on the simulation result. This uncertainty analysis is beyond the scope of the work described in this report but can be evaluated in the future. Ideally, the appropriate complexity of the model will be evaluated according to the model's ability to simulate specific predictions or scenarios of interest. For the set of models described in this report, it is expected that the prediction will be a comparison of future simulation and measurement of age tracer concentrations.

References

Abdallah G., Thoraval A., Sfeir A. and Piguet J. P., 1995. Thermal-convection of fluid in fractured media. *International Journal of Rock Mechanics and Mining Science & Geomechanics Abstracts* 32(5): 481–490.

Ajami N. K., Duan Q. and Sorooshian S., 2007. An integrated hydrologic Bayesian multimodel combination framework: Confronting input, parameter, and model structural uncertainty in hydrologic prediction. *Water Resources Research* 43(1): W01403.

Akaike H., 1974. A new look at the statistical model identification. *IEEE Transactions Automatic Control* 19(06): 716–723.

Allen D. M., Whitfield P. H. and Werner A., 2010. Groundwater level responses in temperate mountainous terrain: Regime classification, and linkages to climate and streamflow. *Hydrological Processes* 24(23): 3392–3412.

Alley W. M., Healy R. W., LaBaugh J. W. and Reilly T. E., 2002. Flow and storage in groundwater systems. *Science* 296(5575): 1985–1990.

Amanambu A. C., Obarein O. A., Mossa J., Li L., Ayeni S. S., Balogun O., Oyebamiji A. and Ochege F. U., 2020. Groundwater system and climate change: Present status and future considerations. *Journal of Hydrology* 589: 125163.

Ataie-Ashtiani B. and Hosseini S. A., 2005. Error analysis of finite difference methods for two-dimensional advection-dispersion-reaction equation. *Advances in Water Resources* 28(8): 793–806.

Ataie-Ashtiani B., Lockington D. A. and Volker R. E., 1996. Numerical correction for finite-difference solution of the advection-dispersion equation with reaction. *Journal of Contaminant Hydrology* 23(1–2): 149–156.

Ataie-Ashtiani B., Lockington D. A. and Volker R. E., 1999. Truncation errors in finite difference models for solute transport equation with first-order reaction. *Journal of Contaminant Hydrology* 35(4): 409–428.

Bai D. M., Qiu T. S. and Li X. B., 2007. The sample entropy and its application in EEG based epilepsy detection. *Journal of Biomedical Engineering* 24(1): 200–205 (in Chinese).

Bai Y. J. and Yin G. D., 2010. Evaluation method and research progress for groundwater quality. *Journal of Water Resources and Water Engineering* 21(03): 115–119 + 123 (in Chinese).

Bak P., Tang C. and Wiesenfeld K., 1987. Self-organized criticality: An explanation of the $1/f$ noise. *Physical Review Letters* 59(4): 381.

Ball W. P., Liu C. X., Xia G. S. and Young D. F., 1997. A diffusion-based interpretation of tetrachloroethene and trichloroethene concentration profiles in a groundwater aquitard. *Water Resources Research* 33(12): 2741–2757.

Bandt C. and Pompe B., 2002. Permutation entropy: A natural complexity measure for time series. *Physical Review Letters* 88(17): 174102.

Barnosky A. D., Brown J. H., Daily G. C., Dirzo R., Ehrlich A. H., Ehrlich P. R., Eronen J. T., Fortelius M., Hadly E. A., Leopold E. B., Mooney H. A., Myers J. P., Naylor R. L., Palumbi S., Stenseth N. C. and Wake M. H., 2014. Introducing the scientific consensus on maintaining humanity's life support systems in the 21st century: Information for policy makers. *The Anthropocene Review* 1(1): 78–109.

Bates J. E. and Shepard H. K., 1993. Measuring complexity using information fluctuation. *Physics Letters A* 172(6): 416–425.

Belytschko T. and Black T., 1999. Elastic crack growth in finite elements with minimal remeshing. *International Journal for Numerical Methods in Engineering* 45(5): 601–620.

Beylich A., Oberholzer H., Schrader S., Hoeper H. and Wilke B., 2010. Evaluation of soil compaction effects on soil biota and soil biological processes in soils. *Soil & Tillage Research* 109(2): 133–143.

Bierkens M. F. P. and Gaast J. W. J. V. D., 1998. Upscaling hydraulic conductivity: Theory and examples from geohydrological studies. *Nutrient Cycling in Agroecosystems* 50(1): 193–207.

Blasone R. S., Vrugt J. A., Madsen H., Rosbjerg D., Robinson B. A. and Zyvoloski G. A., 2008. Generalized likelihood uncertainty estimation (GLUE) using adaptive Markov chain Monte Carlo sampling. *Advances in Water Resources* 31(4): 630–648.

Blois G., Best J. L., Smith G. H. S., and Hardy R. J., 2014. Effect of bed permeability and hyporheic flow on turbulent flow over bed forms. *Geophysical Research Letters* 41(18): 6435–6442.

Bogli A., 1980. *Karst Hydrology and Physical Speleology*. Springer: Berlin.

Bouwer H., 2002. Artificial recharge of groundwater: Hydrogeology and engineering. *Hydrogeology Journal* 10(1): 121–142.

Bravo H. R., Jiang F. and Hunt R. J., 2002. Using groundwater temperature data to constrain parameter estimation in a groundwater flow model of a wetland system. *Water Resources Research* 38(8): 28-1-28-14.

Bredehoeft J., 2005. The conceptualization model problem – surprise. *Hydrogeology Journal* 13: 37–46.

Bremermann H. J. 1962. Optimization through evolution and recombination. *Self-Organizing Systems* 93: 106.

Brunke M. and Gonser T., 1997. The ecological significance of exchange processes between rivers and groundwater. *Freshwater Biology* 37(1): 1–33.

Cao L., 2020. AHP-based analysis of health assessment of small and medium-sized reservoirs in Xinjiang. Xi'an University of Technology (in Chinese).

Carle S. F. and Fogg G. E., 1996. Transition probability-based indicator geostatistics. *Mathematical Geology* 28(4): 453–476.

Chang J. X., Huang Q., Wang Y. M. and Xue X. J., 2002. Water resources evolution direction distinguishing model based on dissipative structure theory and gray relational entropy. *Journal of Water Conservancy* 11(2): 107–112 (in Chinese).

Chapman S. W. and Parker B. L., 2005. Plume persistence due to aquitard back diffusion following dense nonaqueous phase liquid source removal or isolation. *Water Resources Research* 41(12): W12411.

Chen C. X., 2003. Formation mechanism of water level and its determination method in conventional observation wells for three-dimensional groundwater flow. *Earth Science* 28(5): 483–491 (in Chinese).

Chen C. X., 2003. To prevent model loss in reality and improve model accuracy is the key of groundwater numerical modeling. *Hydrogeology & Engineering Geology* 2003(2): 1–5 (in Chinese).

Chen C. X., 2012. Analytical model of groundwater flow in multiaquifer wells. *Hydrogeology & Engineering Geology* 39(5): 1–7 (in Chinese).

Chen C. X. and Hu L. T., 2008. A review of the seepage-pipe coupling model and its application. *Hydrogeology & Engineering Geology* 2008(3): 70–75 (in Chinese).

Chen C. X., Hu L. T. and Wang X. S., 2006. Analysis of steady ground water flow toward wells in a confined-unconfined aquifer. *Ground Water* 44(4): 609–612.

Chen C. X., Lin M. and Cheng J. M., 2011. *Groundwater Dynamics*. Geoscience Press: Beijing (in Chinese).

Chen C. X. and Wan J. W., 2002. A new model of groundwater flowing to horizontal well and the numerical simulation approach. *Earth Science* 27(2): 135–140 (in Chinese).

Chen C. X., Wan J. W. and Zhan H. B., 2003. Theoretical and experimental studies of coupled seepage-pipe flow to a horizontal well. *Journal of Hydrology* 281(1–2): 159–171.

Chen C. X., Wan J. W., Zhan H. B. and Shen Z. Z., 2004. Physical and numerical simulation of seepage-pipe coupling model. *Hydrogeology & Engineering Geology* 2004(1): 1–8 (in Chinese).

Chen C. X., Wang X. S. and Hu L. T., 2007. Emendation of drawdown in pumping wells for numerical modeling of groundwater flow. *Journal of Hydraulic Engineering* 38(4): 481–485 (in Chinese).

Chen J. P., 2001. Towards understanding of dissipative structure deformation process of rock and soil mass. *Journal of Jilin University* 31(3): 288–293 (in Chinese).

Chen M. X. and Ma F. S., 2002. *Groundwater Resources and Environment in China*. Earthquake Press (in Chinese).

Chen N. X. and Huo H. Y., 2008. Water resources system dynamics characteristics of the lower Yellow River. *Journal of Irrigation and Drainage* 27(3): 28–30 (in Chinese).

Chen R. Y., 2008. Dynamic characteristics and genetic analysis of groundwater chemical field in the northern part of Kunming Basin. Kunming University of Science and Technology (in Chinese).

Chen Y., 2013. Evaluation and optimization of hydrologic network based on the theory of information entropy. Wuhan University of Technology (in Chinese).

Chen Y. and Wu J. C., 2005. Effect of the spatial variability of hydraulic conductivity in aquifer on the numerical simulation of groundwater. *Advances in Water Science* 16(4): 482–487 (in Chinese).

Cheng J. M., Chen C. X., Ji M. R. and Sun G. M., 2003. Determination of seaward boundary with three-dimensional density-dependent tidal effect model: By example of coastal aquifers in Jiahe River Basin, Shandong Province. *Earth Science* 28(2): 225–231 (in Chinese).

Cheng J. M., Wan H. E. and Wang J., 2003. Excessive depletion of soil water and regulation and restoration of soil water regime in loess hilly region under *Prunus davidiana* vegetation. *Acta Pedologica Sinica* 40(5): 691–696 (in Chinese).

Cheng S. H., Zhang X. Y., Li M. L., Xie X. M. and Zhao M., 2018. Research on the division of groundwater "red, yellow and blue" based on the critical water level. *Water Science and Engineering Technology* 211(5): 1–4 (in Chinese).

Chi Y. J., 2002. On application and scientific value of dissipative structure theory. *Journal of Sanming College* 19(2): 104–108 (in Chinese).

Christensen T. H., Bjerg P. L., Banwart S. A., Jakobsen R., Heron G. and Albrechtsen H. J., 2000. Characterization of redox conditions in groundwater contaminant plumes. *Journal of Contaminant Hydrology* 45(3–4): 165–241.

Christensen T. H., Kjeldsen P., Bjerg P. L., Jensen D. L., Christensen J. B., Baun A., Albrechtsen H. J. and Heron G., 2001. Biogeochemistry of landfill leachate plumes. *Applied Geochemistry* 16(7–8): 659–718.

Cilliers P., 1998. *Complexity and Postmodernism: Understanding Complex Systems.* Psychology Press: London.

Cornell S. E., Prentice I. C., House J. I. and Downy C. J., 2012. *Understanding the Earth System: Global Change Science for Application.* Cambridge University Press: Cambridge.

Cortes J. E., Munoz L. F., Gonzalez C. A., Nino J. E., Polo A., Suspes A., Siachoque S. C., Hernandez A. and Trujillo H., 2016. Hydrogeochemistry of the formation waters in the San Francisco field, UMV basin, Colombia – A Multivariate Statistical Approach. *Journal of Hydrology* 539: 113–124.

Cui Y. L., Zhao Y. Z., Shao J. L. and He G. P., 2005. On infiltration of the Yellow River in perched section under pumping conditions. *Hydrogeology & Engineering Geology* 2005(1): 57–60 (in Chinese).

Cuthbert M. O., Gleeson T., Moosdorf N., Befus K. M., Schneider A., Hartmann J. and Lehner B., 2019. Global patterns and dynamics of climate-groundwater interactions. *Nature Climate Change* 9(2): 137–141.

Dai H. and Ye M., 2015. Variance-based global sensitivity analysis for multiple scenarios and models with implementation using sparse grid collocation. *Journal of Hydrology* 528: 286–300.

Dasgupta M., Fregoso A., Marzani S. and Salam G. P., 2013. Towards an understanding of jet substructure. *Journal of High Energy Physics* 2013(9): 1–55.

Dehbandi R., Moore F., Keshavarzi B. and Abbasnejad A., 2017. Fluoride hydrogeochemistry and bioavailability in groundwater and soil of an endemic fluorosis belt, central Iran. *Environmental Earth Sciences* 76(4): 177.

Diggle P. and Lophaven S., 2006. Bayesian geostatistical design. *Scandinavian Journal of Statistics* 33(1): 53–64.

Ding F. H., Dai Y., Song H. Y., Wei J. M. and Cha S., 2015. The changing relationship of hydrogeological parameters of Dadianzi well-aquifer system. *Seismology and Geology* 37(4): 982–990 (in Chinese).

Ding W. F. and Li Z. B., 2001. Research trends of soil anti-erodibility. *Technology of Soil and Water Conservation* 2001(1): 36–39 (in Chinese).

Doherty J., 2010. *PEST-Model-Independent Parameter Estimation User Manual* (5th ed.), and addendum. Watermark Numerical Computing: Brisbane, Queensland, Australia.

Dong J., Zhao Y. S., Han R., Liu Y. Y., Li Z. B. and Zong F., 2006. Study on redox zones of landfill leachate plume in subsurface environment. *Environmental Science* 27(9): 1901–1905 (in Chinese).

Dou C. H., Woldt W., Bogardi I. and Dahab M., 1995. Steady state groundwater-flow simulation with imprecise parameters. *Water Resources Research* 31(11): 2709–2719.

Du Y., Deng Y. M., Ma T., Xu Y., Tao Y. Q., Huang Y. W., Liu R. and Wang Y. X., 2020. Enrichment of geogenic ammonium in quaternary alluvial-lacustrine aquifer systems: Evidence from carbon isotopes and DOM characteristics. *Environmental Science & Technology* 54(10): 6104–6114.

Du Y., Ma T., Deng Y. M., Shen S. and Lu Z. J., 2017. Sources and fate of high levels of ammonium in surface water and shallow groundwater of the Jianghan Plain, Central China. *Environmental Science: Processes & Impacts* 19(2): 161–172.

Du Y., Ma T., Deng Y. M., Shen S. and Lu Z. J., 2018. Characterizing groundwater/surface-water interactions in the interior of Jianghan Plain, central China. *Hydrogeology Journal* 26(4): 1047–1059.

Du Z. D., Guo Z. D., Guo Z. M. and Wang Z. M., 2002. Experimental study on effective porosity of aquifer. *Site Investigation Science and Technology* 2002(5): 36–39 (in Chinese).

Dubovikov M. M., Starchenko N. V. and Dubovikov M. S., 2004. Dimension of the minimal cover and local analysis of fractal time series. *Physica A: Statistical Mechanics and Its Applications* 339(3–4): 591–608.

Eberts S. M., Thomas M. A. and Jagucki M. L., 2013. The quality of our nation's waters-factors affecting public supply-well vulnerability to contamination: Understanding observed water quality and anticipating future water quality. *U.S. Geological Survey Circular* 1385: 120.

Ehlers E. and Krafft T., 2006. *Earth System Science in the Anthropocene.* Springer: Berlin.

Elkhoury J. E., Brodsky E. E. and Agnew D. C., 2006. Seismic waves increase permeability. *Nature* 441(7079): 1135–1138.

Emberger F., 1969. Climatique la Tunisia. *Instituto Agronomico perl'Oltremare* 31–52.

Fang J. Q., 1996. Chaos control and synchronization in nonlinear systems and its application prospect. *Progress in Physics* 16(1): 1–23 (in Chinese).

Farmer C. L. and Howison S. D., 2006. The motion of a viscous filament in a porous medium or Hele-Shaw cell: A physical realisation of the Cauchy-Riemann equations. *Applied Mathematics Letters* 19(4): 356–361.

Farrand W. R., Mickelson D. M., Cowan W. R., Goebel J. E., Richmond G. M. and Fullerton D. S., 1984. Quaternary geologic map of the Lake Superior 4 degrees × 6 degrees quadrangle, United States and Canada. *U.S. Geological Survey Miscellaneous Investigations Series Map* 16, 1420.

Feinstein D. T., Fienen M. N., Reeves H. W. and Langevin C. D., 2016. A semi-structured MODFLOW-USG model to evaluate local water sources to wells for decision support. *Ground Water* 54(4): 532–544.

Feinstein D. T., Hunt R. J. and Reeves H. W., 2010. Regional groundwater flow model of the Lake Michigan Basin in support of Great Lakes Basin water availability and use studies. *U.S. Geological Survey Scientific Investigations Report* 51(9): 379.

Feng G. H., Gong H. and Yu S., 2015. Simulation on the groundwater temperature field of groundwater heat pump system. *Procedia Engineering* 121: 1556–1559.

Feynman R. P., Leighton R. B. and Sands M., 2011. *The Feynman Lectures on Physics, Vol. I: The New Millennium Edition: Mainly Mechanics, Radiation, and Heat*. Basic Books: New York.

Fienen M. N., Nolan B. T., Feinstein D. T. and Starn J. J., 2015. Metamodels to bridge the gap between modeling and decision support. *Groundwater* 53(4): 511–512.

Francis H. C., 2000. The significance of microbial processes in hydrogeology and geochemistry. *Hydrogeology Journal* 8(1): 41–46.

Freeze R. A. and Witherspoon P. A., 1967. Theoretical analysis of regional groundwater flow: 2. Effect of water-table configuration and subsurface permeability variation. *Water Resources Research* 3(2): 623–634.

Friedberg R. M., 1958. A learning machine: Part I. *IBM Journal of Research and Development* 2(1): 2–13.

Friedrich C. 1993. *Chaos and Order: The Complex Structure of Living Systems*. VCH: New York.

Gao W. S. and Dong X. B., 2003. Valuation of fragile agriculture ecosystem services in loess hilly-gully region: A case study of Ansai county. *Journal of Natural Resources* 18(2): 182–188 (in Chinese).

Gatland J. R., Santos I. R., Maher D. T., Duncan T. M. and Erler D. V., 2014. Carbon dioxide and methane emissions from an artificially drained coastal wetland during a flood: Implications for wetland global warming potential. *Journal of Geophysical Research: Biogeosciences* 119(8): 1698–1716.

Ge J. Y., Zhou P. and Zhao X., 2008. Research on sleep EEG time-series using nonlinear sample entropy. *Chinese Journal of Electron Devices* 31(3): 972–975 (in Chinese).

Guan X. G., 1999. The dialectics of system's evolution-View of entirety in the theory of dissipative structure. *Journal of Systemic Dialectics* 7(2): 12–15 (in Chinese).

Guo J. Q., 1989. Transfer effects of the observational errors in elastic-releasing coefficient and transmissibility coefficient calculations based on Theis' formula. *Site Investigation Science and Technology* (2): 9–12 (in Chinese).

Guo W., Xiong W., Gao S. S., Hu Z. M., Liu H. L. and Yu R. Z., 2013. Impact of temperature on the isothermal adsorption/desorption characteristics of shale gas. *Petroleum Exploration and Development* 40(4): 514–519.

Han M., Han X. L., Man Y. Y., Li Z. and Hu Y. H., 2005. Multi-dimensional dynamic system model of groundwater resources in karst area. *Shandong Water Resources* (6): 44–45 (in Chinese).

Harbaugh A. W., 2005. *MODFLOW-2005, the US Geological Survey modular ground-water model: The groundwater flow process*. U.S. Department of the Interior, US Geological Survey: Reston, 6–16.

Harbaugh A. W., Banta E. R., Hill M. C. and McDonald M. G., 2000. MODFLOW-2000: The U.S. Geological Survey modular ground-water model: User guide to modularization concepts and the ground-water flow process. *U.S. Geological Survey Open-File Report 2000* 92: 21.

Hassan A. E., Bekhit H. M. and Chapman J. B., 2008. Uncertainty assessment of a stochastic groundwater flow model using GLUE analysis. *Journal of Hydrology* 362(1–2): 89–109.

Hassan A. E., Bekhit H. M. and Chapman J. B., 2009. Using Markov chain Monte Carlo to quantify parameter uncertainty and its effect on predictions of a groundwater flow model. *Environmental Modelling & Software* 24(6): 749–763.

Heilman P., 1975. Effect of added salts on nitrogen release and nitrate levels in forest soils of the Washington coastal area. *Soil Science Society of America Journal* 39(4): 778–782.

Helton J. C. and Oberkampf W. L., 2004. Alternative representations of epistemic uncertainty. *Reliability Engineering & System Safety* 85(1–3): 1–10.

Hendry M. J. and Wassenaar L. I., 2000. Controls on the distribution of major ions in pore waters of a thick surficial aquitard. *Water Resources Research* 36(2): 503–513.

Hendry M. J. and Woodbury A. D., 2007. Clay aquitards as archives of Holocene paleoclimate: $\delta^{18}O$ and thermal profiling. *Groundwater* 45(6): 683–691.

Hill M. C. and Tiedeman C. R., 2007. *Effective Groundwater Model Calibration: With Analysis of Data, Sensitivities, Predictions, and Uncertainty.* John Wiley and Sons: New Jersey.

Hojberg A. L. and Refsgaard J. C., 2005. Model uncertainty: Parameter uncertainty versus conceptual models. *Water Science and Technology* 52(6): 177–186.

Holland J. H., 1962. Outline for a logical theory of adaptive systems. *Journal of the ACM (JACM)* 9(3): 297–314.

Holland J. H., 2014. *Complexity: A Very Short Introduction.* OUP: Oxford.

Hong L. and Xu J. X., 1999. Crises and chaotic transients studied by the generalized cell mapping digraph method. *Physics Letters A* 262(4–5): 361–375.

Hong L. and Xu J. X., 2001. Discontinuous bifurcations of chaotic attractors in forced oscillators by generalized cell mapping digraph (GCMD) method. *International Journal of Bifurcation and Chaos* 11(3): 723–736.

Hsu C. S., 1980. A theory of cell-to-cell mapping dynamical systems. *Journal of Applied Mechanics* 47(4): 931–939.

Hsu C. S., 2013. *Cell-to-Cell Mapping: A Method of Global Analysis for Nonlinear Systems.* Springer Science & Business Media: New York.

Hu L. T. and Chen C. X., 2006a. Application of numerical simulation to the water resources management of the middle reaches of Heihe River Basin. *Geological Science and Technology Information* 25(2): 93–98 (in Chinese).

Hu L. T. and Chen C. X., 2006b. Dynamical simulation for multilayer aquifer system at middle reach of Heihe River Basin. *Journal of System Simulation* 18(7): 1966–1968+1975 (in Chinese).

Hu L. T. and Chen C. X., 2008. Analytical methods for transient flow to a well in a confined-unconfined aquifer. *Ground Water* 46(4): 642–646.

Hu L. T., Chen C. X. and Chen X. H., 2011. Simulation of groundwater flow within observation boreholes for confined aquifers. *Journal of Hydrology* 398(1–2): 101–108.

Hu L. T., Chen C. X., Jiao J. J. and Wang Z. J., 2007. Simulated groundwater interaction with rivers and springs in the Heihe river basin. *Hydrological Processes* 21(20): 2794–2806.

Hu L. T., Chen C., Wang Z. J. and Wang X. S., 2005. Researches on numerical algorithm of groundwater flowline. *Coal Geology & Exploration* 33(5): 33–36 (in Chinese).

Huang W., Wang K., Du H. W., Wang T., Wang S. H., Yangmao Z. M. and Jiang X., 2016. Characteristics of phosphorus sorption at the sediment-water interface in Dongting Lake, a Yangtze-connected lake. *Hydrology Research* 47(S1): 225–237.

Huang X. R., 2005. *Studies on the Methodology of Complexity Sciences.* Tsinghua University: Beijing (in Chinese).

Hunt R. J., Doherty J. and Tonkin M. J., 2007. Are models too simple? Arguments for increased parameterization. *Groundwater* 45(3): 254–262.

Hurvich C. M. and Tsai C. L., 1989. Regression and time series model selection in small samples. *Biometrika* 76(2): 297–307.

Jabakhanji R. and Mohtar R. H., 2015. A peridynamic model of flow in porous media. *Advances in Water Resources* 78: 22–35.

Jaynes E. T., 1957. Information theory and statistical mechanics. *Physical Review*, 106(4): 620.

Jenifer M. A. and Jha M. K., 2017. Comparison of analytic hierarchy process, catastrophe and entropy techniques for evaluating groundwater prospect of hardrock aquifer systems. *Journal of Hydrology* 548: 605–624.

Jiang J. Y., Cao J. F., Li S. and Wang B., 2008. Analysis on the evolution of groundwater system using the theory of dissipative structure. *Research of Soil and Water Conservation* 15(1): 122–124 + 127 (in Chinese).

Jiao J. J., Y. Wang J. A., Cherry X., Wang B., Zhi H., Du H. and Wen D., 2010. Abnormally high ammonium of natural origin in a coastal aquifer-aquitard system in the Pearl River Delta, China. *Environmental Science & Technology* 44(19): 7470–7475.

Jin X., Xu C. Y., Zhang Q. and Singh V. P., 2010. Parameter and modeling uncertainty simulated by GLUE and a formal Bayesian method for a conceptual hydrological model. *Journal of Hydrology* 383(3–4): 147–155.

Juckem P. F., Clark B. R. and Feinstein D. T., 2017. Simulation of groundwater flow in the glacial aquifer system of Northeastern Wisconsin with variable model complexity. U.S. Geological Survey Scientific Investigations Report 2017–5010.

Judd A. and Hovland M., 2009. *Seabed Fluid Flow: The Impact on Geology, Biology and the Marine Environment.* Cambridge University Press: Cambridge.

Kashyap R. L., 1982. Optimal choice of AR and MA parts in autoregressive moving average models. *IEEE Transactions on Pattern Analysis and Machine Intelligence* 4(2): 99–104.

Katz R. W., Parlange M. B. and Naveau P., 2002. Statistics of extremes in hydrology. *Advances in Water Resources* 25(8–12): 1287–1304.

Kaur L., Rishi M. S., Singh G. and Thakur S. N., 2020. Groundwater potential assessment of an alluvial aquifer in Yamuna sub-basin (Panipat region) using remote sensing and GIS techniques in conjunction with analytical hierarchy process (AHP) and catastrophe theory (CT). *Ecological Indicators* 110: 105850.

Kim J. S., Kim S. Y. and Han T. S., 2020. Sensitivity and uncertainty estimation of cement paste properties to microstructural characteristics using FOSM method. *Construction and Building Materials* 242:118159.

Kirshen P. H., 2002. Potential impacts of global warming on groundwater in eastern Massachusetts. *Journal of Water Resources Planning and Management* 128(3): 216–226.

Koltermann C. E. and Gorelick S. M., 1996. Heterogeneity in sedimentary deposits: A review of structure-imitating, process-imitating, and descriptive approaches. *Water Resources Research* 32(9): 2617–2658.

Kong L. N. and Xiang N. P., 2012. Research on rainfall spatial interpolation methods based on ArcGIS. *Geomatics & Spatial Information Technology* 35(3): 123–126 (in Chinese).

Konikow L. F. and Neuzil C. E., 2007. A method to estimate groundwater depletion from confining layers. *Water Resources Research* 43(7): W07417.

Lavigne M., Nastev M. and Lefebvre R., 2010. Numerical simulation of groundwater flow in the Chateauguay River Aquifers. *Canadian Water Resources Journal* 35(4): 469–486.

Lei C., 2018. Study on the risk estimation model of prediction results of mine water inflow. Hefei University of Technology (in Chinese).

Li A. H., Liu H., Geng L. H., Zhong H. P., Jiang B. L., 2009. Review of risk analysis of hydraulic engineering system. *Advances in Water Science* 20(3): 453–459 (in Chinese).

Li D. Q., Jiang J., Liang Y. M., Liu G. B. and Huang J., 1996. Study on water use efficiency of the artificial grassland at Ansai county in the loess hilly region. *Research of Soil and Water Conservation* 3(2): 66–74 (in Chinese).

Li H. and Zhang D. X., 2007. Probabilistic collocation method for flow in porous media: Comparisons with other stochastic methods. *Water Resources Research* 43(9): W09409.

Li M., Chen C. X. and Zhang M. J., 2005. Numerical simulation of three-dimensional groundwater flow of Weigan River Basin, Xinjiang. *Geological Science and Technology Information* 24(1): 74–78 (in Chinese).

Li R. Z., Hong T. Q., Liang Y. Y. and Qian J. Z., 2006. Risk assessment on over-standard pollutant discharged in a pollution source based on unascertained information. *Journal of Wuhan University of Technology* 28(1): 73–76 (in Chinese).

Li X. J., 2000. The alkili-saline land and agricultural sustainable development of the western Songnen Plain in China. *Scientia Geographica Sinica* 20(1): 51–55 (in Chinese).

Li X. L. and Qian J. X., 2003. Studies on artificial fish swarm optimization algorithm based on decomposition and coordination techniques. *Journal of Circuits and Systems* 8(1): 1–6 (in Chinese).

Liang J. W., 2010. Experimental study on soft soil deformation and seepage characteristics with microscopic parameter analysis. South China University of Technology, Guangzhou (in Chinese).

Liao J. F., 1997. Environmental issues in groundwater system evolution: An analysis of Xiaheqing irrigated area groundwater system. *Gansu Water Resources and Hydropower Technology* (1): 28–30 + 51 (in Chinese).

Lineback J. A., Bleuer N. K., Mickelson D. M., Ferrand W. R., Goldthwait R. P., Richmond G. M. and Fullerton D. S., 1983. Quaternary geologic map of the Chicago 4 degrees × 6 degrees quadrangle, United States. *U.S. Geological Survey Miscellaneous Investigations Series Map* 16: 1420.

Liu D., Fu Q. and Zhang J., 2011. Study on complexity of groundwater depth series in well irrigation area of Sanjiang plain based on continuous wavelet transform and fractal theory. *Research of Soil and Water Conservation* 18(2): 116–120 (in Chinese).

Liu M. and Liu D., 2012. Sample entropy based analysis on complicacy of groundwater depth series. *Water Resources and Hydropower Engineering* 43(12): 5–8 (in Chinese).

Liu M., Hou L. J., Xun S. Y., Jiang H. Y., Ou D. N., Yu J. and Wang Q., 2005. The characteristics of NH_4^+—N adsorption on intertidal sediments of the Changjiang Estuary in China. *Acta Oceanologica Sinica* 27(5): 60–66 (in Chinese).

Liu P. G. and Shu L. C., 2008. Uncertainty on numerical simulation of groundwater flow in the riverside well field. *Journal of Jilin University (Earth Science Edition)* 38(4): 639–643.

Liu P. G., Elshall A. S., Ye M., Beerli P., Zeng X. K., Lu D. and Tao Y. Z., 2016. Evaluating marginal likelihood with thermodynamic integration method and comparison with several other numerical methods. *Water Resources Research* 52(2): 734–758.

Liu R., Ma T., Lin C. H., Chen J., Lei K., Liu X. and Qiu W. K., 2020a. Transfer and transformation mechanisms of Fe bound–organic carbon in the aquitard of a lake-wetland system during reclamation. *Environmental Pollution* 263: 114441.

Liu R., Ma T., Qiu W. K., Du Y. and Liu Y. J., 2020b. Effects of Fe oxides on organic carbon variation in the evolution of clayey aquitard and environmental significance. *Science of the Total Environment* 701: 134776.

Liu R., Ma T., Zhang D. T., Lin C. H. and Chen J., 2020c. Spatial distribution and factors influencing the different forms of ammonium in sediments and pore water of the aquitard along the Tongshun River, China. *Environmental Pollution* 266: 115212.

Liu Y., Ma T., Chen J., Xiao C., Liu R., Du Y. and Fendorf S., 2020d. Contribution of clay-aquitard to aquifer iron concentrations and water quality. *Science of the Total Environment* 741: 140061.

Lohmann G., Grosfeld K., Wolf-Gladrow D., Unnithan V., Notholt J. and Wegner A., 2013. *Earth System Science: Bridging the Gaps between Disciplines: Perspectives from a Multi-Disciplinary Helmholtz Research School.* Springer Science & Business Media: Berlin.

Lohmann G., Meggers H., Unnithan V., Wolf-Gladrow D., Notholt J. and Bracher A., 2015. *Towards an Interdisciplinary Approach in Earth System Science.* Springer International Publishing: Berlin.

Long H., Yu Y. J. and Feng Z. Q., 2011. Three-dimensional temperature field simulation of ground heat exchangers with groundwater flow. *Acta Energiae Solaris Sinica* 32(6): 862–867.

Lou Z. H., Cheng J. R. and Jin A. M., 2006. Origin and evolution of the hydrodynamics in sedimentary basins–A case study of the Songliao Basin. *Acta Sedimentica* 24(2): 193–201 (in Chinese).

Luo H. M., Yang P. J., Wang C. J., Yu J. and Mu X., 2015. Lithofacies simulation based on multi-point geostatistics multiple data joint constraints. *Oil Geophysical Prospecting* 50(1): 162–169.

Luther K. H. and Haitjema H. M., 1998. Numerical experiments on the residence time distributions of heterogeneous groundwatersheds. *Journal of Hydrology* 207(1–2): 1–17.

Ma C. Y., Liu Y. T. and Wu J. L., 2013. Simulated flow model of fractured anisotropic media: Permeability and fracture. *Theoretical and Applied Fracture Mechanics* 65: 28–33.

Mackenzie F. T. and Mackenzie J. A., 1995. *Our Changing Planet: An Introduction to Earth System Science and Global Environmental Change.* Prentice Hall: Upper Saddle River, NJ.

Matter J. M., Takahashi T. and Goldberg D., 2007. Experimental evaluation of *in situ* CO_2-water-rock reactions during CO_2 injection in basaltic rocks: Implications for geological CO_2 sequestration. *Geochemistry Geophysics Geosystems* 8(2): Q02001.

Meng L. H., Chen Y. N., Li W. H. and Zhao R. F., 2009. Fuzzy comprehensive evaluation model for water resources carrying capacity in Tarim River basin, Xinjiang, China. *Chinese Geographical Science* 19(1): 89–95.

Meng Q. X., Liu G. B. and Yang Q. K., 2010. Spatial interpolation methods of weather data on Loess Plateau based on GIS. *Research of Soil and Water Conservation* 17(1): 10–14 (in Chinese).

Merritt M. L. and Konikow L. F., 2000. Documentation of a computer program to simulate lake-aquifer interaction using the MODFLOW groundwater flow model and the MOC3D solute-transport model. *U.S. Geological Survey Water Resources Investigations Report* 4167: 146.

Merz B. and Thieken A. H., 2009. Flood risk curves and uncertainty bounds. *Natural Hazards* 51(3): 437–458.

Meyer P. D., Ye M., Neuman S. P., Rockhold M. L., Cantrell K. J. and Nicholson T. J., 2007. *Combined Estimation of Hydrogeologic Conceptual Model, Parameter, and Scenario Uncertainty.* NUREG/CR-6843 Report. Washington, D.C.

Moldrup P., Yamaguchi T., Hansen J. A. and Rolston D. E., 1992. An accurate and numerically stable model for one-dimensional solute transport in soils. *Soil Science* 153(4): 261–273.

Moldrup P., Yamaguchi T., Rolston D. E., Vestergaard K. and Hansen J. A., 1994. Removing numerically induced dispersion from finite-difference models for solute and water transport in unsaturated soils. *Soil Science* 157(3): 153–161.

Montazeri G. H., Ziabakhsh-Ganji Z., Aliabadi H. and Ahanjan A., 2014. The effect of relative permeability on the well testing behavior of naturally fractured lean gas condensate reservoirs. *Petroleum Science and Technology* 32(3): 307–315.

Mossler J. H., 1992. Sedimentary rocks of Dresbachian age (late Cambrian), Hollendale Embayment, southeastern Minnesota. *Minnesota Geological Survey Report of Investigations* 40: 71.

Nettasana T., 2012. Conceptual model uncertainty in the management of the Chi River Basin, Thailand. University of Waterloo.

Neuman S. P., 2003. Maximum likelihood Bayesian averaging of uncertain model predictions. *Stochastic Environmental Research and Risk Assessment* 17(5): 291–305.

Nielsen J. D., 1972. Fixation and release of potassium and ammonium ions in Danish soils. *Plant Soil* 36(1): 71–88.

Notodarmojo S., Ho G. E., Scott W. D. and Davis G. B., 1991. Modeling phosphorus transport in soils and groundwater with 2-consecutive reactions. *Water Research* 25(10): 1205–1216.

Oda M., Takemura T. and Aoki T., 2002. Damage growth and permeability change in triaxial compression tests of Inada granite. *Mechanics of Materials* 34(6): 313–331.

Oliveira P. T. S., Leite M. B., Mattos T., Nearing M. A., Scott R. L., de Oliveira Xavier, R., da Silva Matos D. M. and Wendland E., 2017. Groundwater recharge decrease with increased vegetation density in the Brazilian Cerrado. *Ecohydrology* 10(1): e17591.

Ortega-Guerrero A., Cherry J. A. and Rudolph D. L., 1993. Large-scale aquitard consolidation near Mexico City. *Groundwater* 31(5): 708–718.

Ou G. X., Munoz-Arriola F., Uden D. R., Martin D., Allen C. R. and Shank N., 2018. Climate change implications for irrigation and groundwater in the Republican River Basin, USA. *Climatic Change* 151(2): 303–316.

Owuor S. O., Butterbach-Bahl K., Guzha A. C., Rufino M. C., Pelster D. E., Diaz-Pines E. and Breuer L., 2016. Groundwater recharge rates and surface runoff response to land use and land cover changes in semi-arid environments. *Ecological Processes* 5(1): 1–21.

Packard, N. H., Crutchfield, J. P., Farmer, J. D. and Shaw, R. S., 1980. Geometry from a time series. *Physical Review Letters* 45(9): 712–716.

Pan P. Z., Feng X. T., Huang X. H., Cui Q. and Zhou H., 2009. Coupled THM processes in EDZ of crystalline rocks using an elasto-plastic cellular automaton. *Environmental Geology* 57(6): 1299–1311.

Pan W. N., Kan J. J., Inamdar S., Chen C. and Sparks D., 2016. Dissimilatory microbial iron reduction release DOC (dissolved organic carbon) from carbon-ferrihydrite association. *Soil Biology & Biochemistry* 103: 232–240.

Pan X. P., 2016. Application research of optimized MCMC method in seismic inversion. China University of Petroleum (East China) (in Chinese).

Panday S., Langevin C. D., Niswonger R. G., Ibaraki M. and Hughes J. D., 2013. MODFLOW-USG version 1: An unstructured grid version of MODFLOW for simulating groundwater flow and tightly coupled processes using a control volume finite-difference formulation. *U.S. Geological Survey Techniques and Methods, No. 6-A45.*

Peckhaus V., 2010. The fuzzification of systems: The genesis of fuzzy set theory and its initial applications: Developments up to the 1970s. *ISIS* 101(4): 908–909.

Pelzer G., Tapponnier P., Gaudemer Y., Meyer B., Guo S. M., Yin K. L., Chen Z. T. and Dai H. G., 1988. Offsets of late quaternary morphology, rate of slip, and recurrence of large earthquakes on the Chang Ma fault (Gansu, China). *Journal of Geophysical Research: Solid Earth and Planets* 93(B7): 7793–7812.

Peng T., Chen X. H. and Zhuang C. B., 2009. Analysis on complexity of monthly runoff series based on sample entropy in the Dongjiang River. *Ecology and Environmental Sciences* 18(4): 1379–1382 (in Chinese).

Piao S., Ciais P., Huang Y., Shen Z. H., Peng S. S., Li J. S., Zhou L. P., Liu H. Y., Ma Y. C., Ding Y. H., Friedlingstein P., Liu C. Z., Tan K., Yu Y. Q., Zhang T. Y. and Fang J. Y., 2010. The impacts of climate change on water resources and agriculture in China. *Nature* 467(7311): 43–51.

Poeter E. and Anderson D., 2005. Multimodel ranking and inference in ground water modeling. *Ground Water* 43(4): 597–605.

Pohlmann K., Ye M., Reeves D., Zavarin M. and Chapman J., 2007. *Modeling of groundwater flow and radionuclide transport at the Climax mine sub-CAU, Nevada Test Site*. Desert Research Institute, Nevada System of Higher Education, Reno and Las Vegas, NV.

Potter P. E., Maynard J. B. and Depetris P. J., 2005. *Mud and Mudstones: Introduction and Overview*. Springer Science & Business Media: Berlin, 609.

Prigogine I. and Rysselberghe V. R., 1963. Introduction to Thermodynamics of Irreversible Processes. *Journal of the Electrochemical Society* 110(4): 97C.

Pryshlak T. T., Sawyer A. H., Stonedahl S. H. and Soltanian M. R., 2015. Multiscale hyporheic exchange through strongly heterogeneous sediments. *Water Resources Research* 51(11): 9127–9140.

Qian H. and Hu J. G., 1996. Bogli mixed corrosion theory and its problems in practical application. *Carsologica Sinica* 15(04): 367–375 (in Chinese).

Qian H., Lian J. and Dou Y., 2007. Mixing effects of groundwater on $CaCO_3$ dissolution and precipitation. *Journal of Earth Sciences and Environment* 29(01): 55–65 (in Chinese).

Qian H., Ma Z. Y. and Li P. Y., 2012. *Hydrogeochemistry*. Geology Press: Beijing (in Chinese).

Qian X. S., 2000. *Macro Architecture and Micro Architecture*. Hangzhou Press (in Chinese).

Qian X. S., Yu J. Y. and Dai R. W., 1990. A new discipline of science: The study of open complex giant system and its methodology. *Chinese Journal of Nature* 01(1): 3–10 (in Chinese).

Qin B. H., 1994. Application of some new theories to geology – an introduction to dissipative structure chaos and fractal theories. *Hunan Geology* 4: 241–249 (in Chinese).

Qin R. G., Cao G. Z. and Wu Y. Q., 2014. Review of the study of groundwater flow and solute transport in heterogeneous aquifer. *Progress in Earth Science* 29(01): 30–41 (in Chinese)

Qiu M. D., Su Q. S. and Zhao Y. H., 1989. Research on the coefficients of water release and storage in the aquifer. *Journal of Changchun University of Earth Science* 19(2): 191–198 (in Chinese).

Rechenberg I., 1965. Cybernetic solution path of an experimental problem. *Royal Aircraft Establishment Library Translation*.

Refsgaard J. C., Christensen S., Sonnenborg T. O., Seifert D., Hojberg A. L. and Troldborg L., 2012. Review of strategies for handling geological uncertainty in groundwater flow and transport modeling. *Advances in Water Resources* 36: 36–50.

Renard B., Kavetski D., Kuczera G., Thyer M. and Franks S. W., 2010. Understanding predictive uncertainty in hydrologic modeling: The challenge of identifying input and structural errors. *Water Resources Research* 46(5): W05521.

Richman J. S. and Moorman J. R., 2000. Physiological time-series analysis using approximate entropy and sample entropy. *American Journal of Physiology Heart & Circulatory Physiology* 278(6): H2039–H2049.

Rissanen J., 1978. Modeling by shortest data description. *Automatica* 14(5): 465–471.

Rojas R., Batelaan O., Feyen L. and Dassargues A., 2010. Assessment of conceptual model uncertainty for the regional aquifer Pampa del Tamarugal-North Chile. *Hydrology and Earth System Sciences* 14(2): 171–192.

Rojas R., Feyen L. and Dassargues A., 2008. Conceptual model uncertainty in groundwater modeling: Combining generalized likelihood uncertainty estimation and Bayesian model averaging. *Water Resources Research* 44(12): W12418.

Rutqvist J., Freifeld B., Min K., Elsworth D. and Tsang Y., 2008. Analysis of thermally induced changes in fractured rock permeability during 8 years of heating and cooling at the Yucca Mountain Drift Scale Test. *International Journal of Rock Mechanics and Mining Sciences* 45(8): 1373–1389.

Rycroft R. W. and Kash D. E., 1999. *The Complexity Challenge: Technological Innovation for the 21st Century.* Cengage Learning EMEA: Stamford.

Saad D. A., 1997. Effects of landuse and geohydrology on the quality of shallow ground water in two agricultural areas in the western Lake Michigan drainages. Wisconsin. U.S. Geological Survey Water-Resources Investigations Report 96–4292, 77 p.

Saad D. A. and Thorstenson D. C., 1998. Flow and geochemistry along shallow groundwater flowpaths in an agricultural area in southeastern Wisconsin (Vol. 98, No. 4179). US Department of the Interior, US Geological Survey.

Saad D. A., 2008. Agriculture-related trends in groundwater quality of the glacial deposits aquifer, central Wisconsin. *Journal of Environmental Quality* 37(5): 209–225.

Salehin M., Packman A. I. and Paradis M., 2004. Hyporheic exchange with heterogeneous streambeds: Laboratory experiments and modeling. *Water Resources Research* 40: W11504.

Samani S. and Moghaddam A. A., 2015. Hydrogeochemical characteristics and origin of salinity in Ajabshir aquifer, East Azerbaijan, Iran. *Quarterly Journal of Engineering Geology and Hydrogeology* 48(3–4): 175–189.

Samani S., Moghaddam A. A. and Ye M., 2018. Investigating the effect of complexity on groundwater flow modeling uncertainty. *Stochastic Environmental Research and Risk Assessment* 32(3): 643–659.

Sawyer A. H. and Cardenas M. B., 2009. Hyporheic flow and residence time distributions in heterogeneous cross-bedded sediment. *Water Resources Research* 45(8): W08406.

Schellnhuber H. J., 1999. 'Earth system' analysis and the second Copernican revolution. *Nature* 402: C19–C23.

Schellnhuber H. J. and Wenzel V., 2012. *Earth System Analysis: Integrating Science for Sustainability.* Springer Science & Business Media.

Schulz K., Huwe B. and Peiffer S., 1999. Parameter uncertainty in chemical equilibrium calculations using fuzzy set theory. *Journal of Hydrology* 217(1–2): 119–134.

Schwarz G., 1978. Estimating the dimension of a model. *Annals of Statistics* 6(2): 461–464.

Scibek J. and Allen D. M., 2006. Modeled impacts of predicted climate change on recharge and groundwater levels. *Water Resources Research* 42(11): W11405.

Selroos J. O., Walker D. D., Strom A., Gylling B. and Follin S., 2002. Comparison of alternative modelling approaches for groundwater flow in fractured rock. *Journal of Hydrology* 257(1–4): 174–188.

Shen W., 2001. The theory of self-organization and dissipation structure and its geo-application. *Earth and Environment* 29(3): 1–7 (in Chinese).

Shou Y. D., 2017. Peridynamic numerical simulation of thermo-hydro-mechanical coupled problems in crack-weakened rock. Chongqing University (in Chinese).

Silva L. C. R. and Lambers H, 2021. Soil-plant-atmosphere interactions: Structure, function, and predictive scaling for climate change mitigation. *Plant Soil* 461(1–2): 5–27.

Singh A., Mishra S. and Ruskauff G., 2010. Model averaging techniques for quantifying conceptual model uncertainty. *Ground Water* 48(5): 701–715.

Singh S. K., 2006. Identification of aquifer parameters from residual drawdowns: An optimization approach. *Hydrological Sciences Journal* 51(6): 1139–1148.

Sinha R. K. and Geiser J., 2007. Error estimates for finite volume element methods for convection-diffusion-reaction equations. *Applied Numerical Mathematics* 57(1): 59–72.

Smith R., Knight R. and Fendorf S., 2018. Overpumping leads to California groundwater arsenic threat. *Nature Communications* 9: 2089.

Soller D. R., Packard P. H. and Garrity C. P., 2012. Database for USGS Map I-1970: Map showing the thickness and character of Quaternary sediments in the glaciated United States east of the Rocky Mountains. *U.S. Geological Survey Data Series* 656.

Srinu S. and Mishra A. K., 2016. Cooperative sensing based on permutation entropy with adaptive thresholding technique for cognitive radio networks. *IET Science, Measurement & Technology* 10(8): 934–942.

Sruthi K. V. and Hyun-Su K., 2016. Effect of numerical truncation error on implicit finite difference methods in groundwater transport models. *Current Science* 111(4): 694–699.

Su W. S., Wang F. T., Zhu H., Guo Z. G., Zhang Z. X. and Zhang H. Y., 2011. Feature extraction of rolling element bearing fault using wavelet packet sample entropy. *Journal of Vibration, Measurement & Diagnosis* 31(2): 162–166 (in Chinese).

Sun C. Z. and Liu Y. Y., 2009. Construction of evaluation index system for groundwater ecosystem health assessment. *Acta Ecologica Sinica* 29(10): 5665–5674 (in Chinese).

Sun N. Z., 2013. *Inverse Problems in Groundwater Modeling*. Springer Science & Business Media: Berlin.

Sun X. L. and Zhang Z. Y., 2004. *Ten Relationships in Scientific Methodology*. Academia Press: Shanghai (in Chinese).

Tang M. G. and Wang H. J., 1986. Significance and application of study on the composition of rock-soil pore solution. *Reconnaissance Science and Technology* 5: 1–6 (in Chinese).

Taufiq A., Hosono T., Ide K., Kagabu M., Iskandar I., Effendi A. J., Hutasoit L. M. and Shimada J., 2018. Impact of excessive groundwater pumping on rejuvenation processes in the Bandung basin (Indonesia) as determined by hydrogeochemistry and modeling. *Hydrogeology Journal* 26(4): 1263–1279.

Taylor C. A. and Stefan H. G., 2009. Shallow groundwater temperature response to climate change and urbanization. *Journal of Hydrology* 375(3–4): 601–612.

Terano T., Kiyoji Asai K., and Sugeno M., 1992. Fuzzy Systems Theory and Its Applications. Academic Press Professional, Inc., USA.

Tesoriero A. J., Duff J. H., Saad D. A., Spahr N. E. and Wolock D. M., 2013. Vulnerability of streams to legacy nitrate sources. *Environmental Science and Technology* 47(8): 3623–3629.

Tiwari A. K., Maio M. D., Singh P. K. and Singh A. K., 2016. Hydrogeochemical characterization and groundwater quality assessment in a coal mining area, India. *Arabian Journal of Geosciences* 9(3): 1–17.

Tongue B. and Gu K., 1988. Interpolated cell mapping of dynamical systems. *Journal of Applied Mechanics* 55(2): 461–466.

Troldborg L., Refsgaard J. C., Jensen K. H. and Engesgaard P., 2007. The importance of alternative conceptual models for simulation of concentrations in a multi-aquifer system. *Hydrogeology Journal* 15(5): 843–860.

Tsang C., Stephansson O., Jing L. and Kautsky F., 2009. Decovalex project: From 1992 to 2007. *Environmental Geology* 57(6): 1221–1237.

Turner D, 2018. *The Green Marble: Earth System Science and Global Sustainability*. Columbia University Press: New York.

Vrugt J. A., Diks C. G. H. and Clark M. P., 2008. Ensemble Bayesian model averaging using Markov chain Monte Carlo sampling. *Environmental Fluid Mechanics* 8(5): 579–595.

Waber H. N. and Smellie J. A. T., 2008. Characterisation of pore water in crystalline rocks. *Applied Geochemistry* 23(7): 1834–1861.

Wallis P. J. and Ison R. L., 2011. Appreciating institutional complexity in water governance dynamics: A case from the Murray-Darling Basin, Australia. *Water Resources Management* 25(15): 4081–4097.

Wan J. W., Huang K. and Chen C. X., 2013. Reassessing Darcy' law on water flow in porous media. *Earth Science* 38(6): 1327–1330 (in Chinese).

Wang D. C., Zhang R. Q., Shi Y. H., Xu S. Z., Yu Q. C. and Liang X., 1995. *Foundation of Hydrogeology*. Geological Publishing House: Beijing (in Chinese).

Wang G. Y., 1990. Unascertained information and its mathematical treatment. *Journal of Harbin University of Civil Engineering and Architecture* 23(4): 1–9 (in Chinese).

Wang J., 2005. Heavy-metal elements transfer, distribution and ecological environment impact by mining in the Xiangsi river area, Tongling. Hefei University of Technology (in Chinese).

Wang J. J. and Fan Z. R., 2017. Influence of aquifer heterogeneity on Monte Carlo simulation results of groundwater. *Water Resources Protection* 33(01): 46–51 (in Chinese).

Wang J. J., Lu X. Z., Qiu X. Z. and Liang Y., 2013. Experimental studies on influence factors of permeability coefficients of coarse-grained soil. *Hydro-Science and Engineering* 142(6): 16–20 (in Chinese).

Wang J. Q., Liu W. R., Qian J. Z. and Sun S. Q., 2002. Grey model of groundwater quality assessment based on single factor contaminant index. *Journal of Hefei Polytechnic University (Natural Edition)* 25(5): 697–702 (in Chinese).

Wang S. J. and Hsu K. C., 2013. Dynamic interactions of groundwater flow and soil deformation in randomly heterogeneous porous media. *Journal of Hydrology* 499: 50–60.

Wang S. R., Jin X. C., Zhao H. C., Pang Y., Zhou X. N. and Chu J. Z., 2005. Phosphate adsorption characteristics onto the sediments from shallow lakes in the middle and lower reaches of the Yangtze River. *Environmental Science* 26(3): 38–43.

Wang X. S. and Chen C. X., 2002. Analysis of modified Theis model on well flow: Considering bending of the confining stratum. *Earth Science* 27(2): 199–202 (in Chinese).

Wang X. S., Chen C. X. and Jiao J. J., 2003. Theory on coupling of radial flow in a confined aquifer and bending of confining unit. *Earth Science* 28(5): 545–550 (in Chinese).

Wang Y., Jiao J. J., Cherry J. A. and Lee C. M., 2013. Contribution of the aquitard to the regional groundwater hydrochemistry of the underlying confined aquifer in the Pearl River Delta, China. *Science of the Total Environment* 461–462: 663–671.

Wang Y., Jiao J. J., Zhang K. and Zhou Y., 2016. Enrichment and mechanisms of heavy metal mobility in a coastal quaternary groundwater system of the Pearl River Delta, China. *Science of the Total Environment* 545–546: 493–502.

Wang Y. X., Zhu Y. H. and Zhang Z. H., 2003. The background, problems and development strategy of earth system science. *Journal of China University of Geosciences (Social Sciences Edition)* 3(1): 9–12 (in Chinese).

Wang Z., Zhang L., Dang X. L. and Zhang Y. L., 2012. Effect of the freezing-thawing on kinetics of adsorption-desorption of the soil cadmium. *Acta Scientiae Circumstantiae* 32(3): 721–725 (in Chinese).

Wang Z. B., Shen L. F. and Xu Z. M., 2016. Hydrochemical characteristics and their implication for the water-rock/soil interaction in the Touzhai Landslide. *Hydrogeology & Engineering Geology* 43(1): 111–116, 123 (in Chinese).

Wang Z. L., Geng Y. F. and Jin S., 2005. Improvement of water level calculation of rivers in mountainous district. *Journal of Dalian University of Technology* 45(03): 433–437 (in Chinese).

Weill S., Delay F., Pan Y. and Ackerer P., 2017. A low-dimensional subsurface model for saturated and unsaturated flow processes: Ability to address heterogeneity. *Computational Geosciences* 21(2): 301–304.

Wen D. G., Zhang E. Y., Tang Z. H., Lin L. J., Wan L., Chen C. X., Shao J. L., Zhao Y. S., Wang W. K., Qian H. and Li W. P., 2011. Advances in basin-scale groundwater modelling in China. IAHS-AISH publication, 341: 63–69.

Wen X. H. and Gómez-Hernández J. J., 1996. Upscaling hydraulic conductivities in heterogeneous media: An overview. *Journal of Hydrology* 183(1–2): ix–xxxii.

Westenbroek S. M., Kelson V. A., Dripps W. R., Hunt R. J. and Bradbury K. R., 2010. SWB: A modified Thornthwaite-Mather soil-water-balance code for estimating groundwater recharge. *US Geological Survey Techniques and Methods*, book 6, chap A31.

White W. B., 1988. *Geomorphology and Hydrology of Karst Terrains.* Oxford University Press: New York.

White W. B., 2002. Karst hydrology: Recent developments and open questions. *Engineering Geology* 65(2–3): 85–105.

Wilson J. L., 1997. Removal of aqueous phase dissolved contamination: Non-chemically enhanced pump-and-treat. *Subsurface Restoration* 17: 271–285.

Winz I., Brierley G. and Trowsdale S., 2009. The use of system dynamics simulation in water resources management. *Water Resources Management* 23(7): 1301–1323.

Wood J. R. and Hewett T. A., 1982. Fluid convection and mass transfer in porous sandstones—A theoretical-model. *Geochimica et Cosmochimica Acta* 46(10): 1707–1713.

Wu J. C. and Zeng X. K., 2013. Review of the uncertainty analysis of groundwater numerical simulation. *Chinese Science Bulletin* 58(25): 3044–3052.

Wu J. C., Lu L. and Tang T., 2011. Bayesian analysis for uncertainty and risk in a groundwater numerical model's predictions. *Human and Ecological Risk Assessment: An International Journal* 17(6): 1310–1331.

Wu L. P., Zhu C. J. and Li S., 2012. The gray analysis of Xiaonanhai spring water table. *Groundwater* 34(2): 66–68 (in Chinese).

Wu T., 2001. Rise of the paradigm for complexity. *Science, Technology and Dialectics* 18(6): 20–24 (in Chinese).

Wu T., 2002. On relationships of complexity and randomicity. *Journal of Dialectics of Nature* 24(2): 18–23 (in Chinese).

Wu X. M., Chen C. X., Shi S. S. and Li Z. H., 2003. Three-dimensional numerical simulation of groundwater system in Ejina Basin, Heihe River, Northwestern China. *Earth Science* 28(5): 527–532 (in Chinese).

Xi D. Y., Cheng J. Y. and Xi J., 1998. Research on the heating activation mechanism of viscoelastic relaxation in fluid-saturated sandstone. *Oil Geophysical Prospecting* 33(3): 348–354 (in Chinese).

Xin W. Y., Zhang X. D., Xu S. T. and Wang L. J., 2019. Brief discussions on factors controlling permeability coefficient measurements in aquifer. *Inner Mongolia Water Resources* (12): 20–22 (in Chinese).

Xing L. N., 2012. Groundwater hydrochemical characteristics and hydrogeochemical processes approximately along flow paths in the North China Plain. China University of Geosciences, Beijing (in Chinese).

Xu H. L., Xiao G. Q., Li H., 2002. Evolution of quaternary groundwater system in Hebei Plain under human activities. *Geological Science and Technology Information* 21(1): 7–13 (in Chinese).

Xu J. H., 2002. *Mathematical Methods in Modern Geography*. Higher Education Press: Beijing, 392–417 (in Chinese).

Xu W. D., 2016. Study on the relationship between groundwater and vegetation. *Energy and Energy Conservation* 127(4): 97–98 (in Chinese).

Xu X. W., Li B. W. and Wang X. J., 2006. Progress in study on irrigation practice with saline groundwater on sandlands of Taklimakan Desert Hinterland. *Chinese Science Bulletin* 51 (Suppl 1): 133–136.

Xue Y. Q. and Wu J. C., 1999. Problems of groundwater modeling in China facing 21st century. *Hydrogeology and Engineering Geology* (5): 1–3 (in Chinese).

Yan B. R., Gao Y. S. and Zhang S., 1987. A study of microbial activity mechanism on the groundwater quality evolution and injection well clogging during injecting waste water of air conditioning. *Collected Works of Institute of Hydrogeology and Engineering Geology, Chinese Academy of Geological Sciences* (3): 58–95 (in Chinese).

Yan R., Wang G. and Shi Z., 2016. Sensitivity of hydraulic properties to dynamic strain within a fault damage zone. *Journal of Hydrology* 543(Part B): 721–728.

Yan R. Q. and Gao R. X., 2007. Approximate entropy as a diagnostic tool for machine health monitoring. *Mechanical Systems and Signal Processing* 21(2): 824–839.

Yan T. T. and Wu J. F., 2006. Impacts of the spatial variation of hydraulic conductivity on the transport fate of contaminant plume. *Advances in Water Science* 17(1): 29–36 (in Chinese).

Yan Y. N., 2018. Research on Heihe groundwater level forecast based on grey theory and machine learning. Lanzhou University (in Chinese).

Yan Z. W., Liu H. L. and Tao Z. T., 2011. Temperature effect on carbonic acid balance in water. *Carsologica Sinica* 30(2): 128–131 (in Chinese).

Yang J., Zhang D. M., Jia L. P. and Zhao H., 1998. On the development of the concept of entropy to the theory of dissipation structure. *Journal of Hebei University of Geology* 21(5): 500–506 (in Chinese).

Yang P. H., Yuan D. X., Yuan W. H., Kuang Y. L., Jia P. and He Q. F., 2010. Formations of groundwater hydrogeochemistry in a karst system during storm events as revealed by PCA. *Chinese Science Bulletin* 55(14): 1412–1422.

Ye M., Meyer P. D. and Neuman S. P., 2008. On model selection criteria in multi model analysis. *Water Resources Research* 44(3): W03428.

Ye M., Neuman S. P. and Meyer P. D., 2004. Maximum likelihood Bayesian averaging of spatial variability models in unsaturated fractured tuff. *Water Resources Research* 40(5): W05113.

You X. J., Liu S. G., Dai C. M., Guo Y. P., Zhong G. H. and Duan Y. P., 2020. Contaminant occurrence and migration between high- and low-permeability zones in groundwater systems: A review. *Science of the Total Environment* 743: 140703.

Yu C. W., 1998. Complexity and self-organized criticality of solid earth system. *Geological Journal of China Universities* 4(4): 361–368 (in Chinese).

Yu K., 2016. The sources and influences of dissolved organic matter on temporal variations of groundwater arsenic concentrations: A case study in Jianghan Plain. China University of Geosciences (Wuhan) (in Chinese).

Yu X. W. and Liu D., 2012. Complexity measure for regional groundwater system based on Kolmogorov Entropy and its analysis of influence on the model prediction and precision. *China Rural Water and Hydropower* (8): 65–69+71 (in Chinese).

Yu M., Liu D. and Bazimenyera J. D. D., 2013. Diagnostic complexity of regional groundwater resources system based on time series fractal dimension and artificial fish swarm algorithm. *Water Resources Management* 27(7): 1897–1911.

Yuan W. Z., 2017. Biogeochemical process of Fe and Mn during river bank infiltration affected by groundwater exploiting. Jilin University (in Chinese).

Zeng X. K., 2012. The uncertainty analysis and assessment of groundwater flow numerical simulation. Nanjing University (in Chinese).

Zeng X. K., Wang D., Wu J. C. and Chen X., 2013. Reliability analysis of the groundwater conceptual model. *Human and Ecological Risk Assessment: An International Journal* 19(2): 515–525.

Zhang C. C., Shao J. L., Li C. J. and Cui Y. L., 2003. Eco-environmental effects on groundwater and its eco-environmental index. *Hydrogeology and Engineering Geology* (3): 6–10 (in Chinese).

Zhang G. H., 2009. *Theory and methodology of regional groundwater function and sustainable utilization assessment in China.* Geological Publishing House: Beijing (in Chinese).

Zhang G. H., Fei Y. H., Liu K. Y., 2004. *Groundwater Evolution and Countermeasures in Haihe Plain.* Science Press: Beijing (in Chinese).

Zhang G. H., Nie Z. L., Chen Z. Y., Lai Q. B. and Wang J. Z., 2001. Evolutionary process of hydrologic cycle and periodicity of groundwater change in North China Plain since Holocene. *Acta Geoscience Sinica* 22(4): 293–297 (in Chinese).

Zhang H. T., Xu G. Q., Chen X. Q. and Mabaire A., 2019. Hydrogeochemical evolution of multilayer aquifers in a massive coalfield. *Environmental Earth Sciences* 78(24): 675.

Zhang J. W., Liang X., Jin M. G., Ma T., Deng Y. M. and Ma B., 2019. Identifying the groundwater flow systems in a condensed river-network interfluve between the Han River and Yangtze River (China) using hydrogeochemical indicators. *Hydrogeology Journal* 27(7): 2415–2430.

Zhang J. W., Lu W. X., Qu Y. G. and An Y. K., 2018. Uncertainty analysis of surface water and groundwater coupling simulation model based on Monte Carlo method. *Journal of Water Resources* 49(10): 1254–1264 (in Chinese).

Zhang J. W., Ma T., Feng L. and Yan Y. N., 2015. Conceptual model of redox zonation in high arsenic groundwater system under the effect of microorganism. *Geological Science and Technology Information* 34(5): 153–159 (in Chinese).

Zhang M. J., Men G. F. and Chen C. X., 2004. Three-dimensional digital simulating of groundwater flow of Weigan river drainage area. *Xinjiang Geology* 22(3): 238–243 (in Chinese).

Zhang P., 2009. Nonlinear study on surrounding rock stability of Baihe Tunnel based on wavelet theory. Jilin University, Changchun (in Chinese).

Zhang R. Q., Liang X., Jin M. G., Wan L. and Yu Q. C., 2011. *General Hydrogeology*. Geology Press: Beijing (in Chinese).

Zhang S., Zhang C. Y., Li Y., Li Z. H. and Zhang M., 2005. Geomicrobial geochemical processes: Significance and prospects. *Geological Bulletin of China* 24(10–11): 1027–1031 (in Chinese).

Zhang Y. J. and Liu D., 2017. Groundwater system complexity measurement in urbanized area based on permutation entropy. *Water Saving Irrigation* (6): 50–54 (in Chinese).

Zhang Z. G., Tang Z. L. and Luo Z. J., 2007. On flow field analysis of hydrodynamic field in hydrogeological conceptual model. *Jiangsu Geology* 31(2): 143–146 (in Chinese).

Zhang Z. Y., 2009. Error estimates of finite volume element method for the pollution in groundwater flow. *Numerical Methods for Partial Differential Equations* 25(2): 259–274.

Zhao J., 2010. Comparison of interpolation method of ArcGIS and its application in soil pollution investigation of Yunnan. *Environmental Science Survey* 29(S1): 85–87 (in Chinese).

Zhao J. G., Jia H. Y., Zhan Y. L., Xiang Z. Y., Zheng S. X. and Bi K. M., 2020. Combination of LS-SVM algorithm and JC method for fragility analysis of deep-water high piers subjected to near-field ground motions. *Structures* (24): 282–295.

Zhao Y. Z., Shao J. L., Yan Z. P., Cui Y. L., Jiao H. J. and He G. P., 2003. A preliminary study on river recharge area of groundwater around the Yellow River. *Yellow River* 25(1): 3–5 (in Chinese).

Zheng F. L. and Gao X. T., 2000. *Soil Erosion Processes and Modeling at Loessial Hillslope.* Shanxi People's Press: Xian (in Chinese).

Zhou X. J., 2004. Some cognitions on earth system science. *Advances in Earth Science* 19(4): 513–515 (in Chinese).

Zhou X. Z., Gao Q., Chen X. L., Yu M. and Zhao X. W., 2013. Numerically simulating the thermal behaviors in groundwater wells of groundwater heat pump. *Energy* 61: 240–247.

Zhu B., Chen S., You X., Peng K. and Zhang X. W., 2002. Soil fertility restoration on degraded upland of purple soil. *Acta Pedologica Sinica* 39(5): 743–749 (in Chinese).

Zhu Z. C., 1999. *Structural Geology.* China University of Geosciences Press: Wuhan (in Chinese).

Zuo Q. T. and Ma J. X., 1994. Indeterminate information in groundwater systems and its processing methods. *Hydrogeology & Engineering Geology* 21(5): 41–43 (in Chinese).

Index